INTERNATIONAL DEVELOPMENT IN FOCUS

Digital Senegal for Inclusive Growth

Technological Transformation
for Better and More Jobs

MARCIO CRUZ, MARK A. DUTZ, AND CARLOS RODRÍGUEZ-CASTELÁN

 **WORLD BANK GROUP**

Contents

Boxes

Figures

Tables

Foreword

Senegal recently adjusted the Priority Action Plan for phase II of the Emerging Senegal Plan (PAP2A), which, in its strategic framework, underlines the imperative of reducing the digital divide among the lessons learned. Indeed, digital technologies, the uses of which have become essential nowadays—including to help meet the challenges posed by the COVID-19 crisis—are placed at the heart of the priorities of PAP2A. Most of PAP2A's activities should be based on digital technologies, particularly as engines of growth and job creation.

This ambition to make digital transformation one of the catalysts of more inclusive growth, with better jobs for all, is reaffirmed in the Digital Senegal 2025 (SN2025) strategy, which has the vision of "digital for all and for all uses in 2025 in Senegal, with a dynamic and innovative private sector in an efficient ecosystem." In this perspective, the state and the operators of the sector, as well as new digital entrepreneurs, intend to continue efforts to ensure the coverage of the territory by optical fiber to make available high- and very-high-speed internet services, fixed and mobile, at affordable prices and to accelerate the adoption and intensive use of digital technologies in all sectors.

This report explores possible solutions for a more intensive use of digital technology, especially by small and medium enterprises, to increase their productivity and create more quality jobs. In this context, the report undoubtedly provides insights that will contribute to helping women and young people in particular to gain access to decent work and therefore reduce their exposure to poverty.

Thus, appropriate use of this report will make it possible to succeed in the challenges of digital transformation, especially in the context of a relatively young population that is more open to innovation and change. The Jobs and Economic Transformation and Recovery (JET) program promotes this transformation, aiming to accelerate the competitiveness of micro, small, and medium-size enterprises, formal as well as informal, and job creation. The JET program does this through facilitation of access to technologies and skills, and promotion of competitiveness at the level of value chains, supported by public-private partnerships and better access to financing.

Amadou Hott
Minister of Economy, Planning and
International Cooperation
Senegal

Foreword

The world is undergoing a digital revolution that is reshaping how and where better and more jobs are created—centered around investments in data-driven technologies such as data sensors and platforms. To achieve their potential, these investments need complementary outlays such as in power, tractors, and irrigation systems, and in soft technologies such as better management practices and innovative public-private partnerships.

This book provides an assessment of Senegal's technological capabilities and proposes a set of public policies to achieve more prosperous and inclusive growth. Its focus is on understanding how technological transformation can benefit enterprises and households. The book is centered on new measures of technologies actually used by enterprises. These data are essential to understand the link between technology and jobs. The book argues that better-paying and more jobs are created by increasingly productive and competitive firms, including informal household enterprises, enabled by using better technologies and capabilities. It is through the increased production that results from better technologies that these jobs are generated.

The main findings of this book are that enterprises using better technologies indeed generate better and more jobs, including for lower-skilled workers, and that households covered by mobile internet have better jobs with higher earnings and are associated with lower poverty. Encouragingly, it finds that Senegal has had increased use of digital technologies since the onset of COVID-19. However, large technological gaps remain between Senegal and comparator countries. Within Senegal, divides in terms of access to and use of digital technologies remain between men and women, richer and poorer households, urban and rural people, and higher- and lower-skilled workers. Of great concern, digital divides are increasing between larger enterprises that are investing at a higher rate in better technologies and smaller enterprises that are not. Unaddressed, these gaps will likely worsen over time.

The book recommends that the government should build on its existing policies to ensure more rapid availability of affordable digital infrastructure for all. However, this is not enough. The book also recommends that the government should implement targeted incentives to promote use of better technologies as well as policies to narrow deepening digital divides across enterprises and households. Promisingly, the emerging findings of this book have served as analytical underpinnings to help design a national government program to accelerate competitiveness of small and medium enterprises and job creation supported by the World Bank. It includes public policy incentives to support adoption of technologies and capabilities by firms through matching grants coupled with increased access to finance through partial credit guarantees and equity investments. The national program also includes policies to reinforce competitiveness of value chains focused on fixing coordination problems and promoting shared assets and access to markets, as well as the structuring of public-private partnerships for needed investments in infrastructure and services—mirroring the recommendations of this book. Indeed, the technology measures developed to understand differences across firms now also serve as credible indicators of whether the expected program outcomes will be met. Other complementary actions including direct private investments in medium-size firms by the International Finance Corporation and political risk insurance support to foreign direct investors by the Multilateral Investment Guarantee Agency would ideally complement and complete World Bank Group support to Senegal in this area. The goal is to improve the competitiveness of selected value chains, enterprise capabilities, and private sector investments so that technological transformation indeed translates into more inclusive growth in Senegal.

Ousmane Diagana
Vice President
Western and Central Africa Region
The World Bank

Acknowledgments

This book was prepared as part of the World Bank analytical and advisory services task "Senegal Digital Sources of Growth Study: Productivity and Inclusion Opportunities from Adoption of Digital Technologies (DTs)" (project ID: P168247), and as part of the joint Africa Chief Economist (AFRCE)–Digital Development Research Program on Digital Transformation for Africa (project ID: P170151). It was prepared by a joint AFRCE, Finance, Competitiveness, and Innovation Global Practice and Poverty and Equity Global Practice team led by Mark A. Dutz, Marcio Cruz, and Carlos Rodríguez-Castelán, with World Bank budget overseen by Finance, Competitiveness, and Innovation Practice Manager Consolate Rusagara. The core World Bank team included Laurent Corthay, Laurent Gonnet, Aneliya Muller, Meriem Ait Ali Slimane, and Mariana Vijil. This work was produced in close coordination with the government of Senegal and it received grant support from the Competitive Industries and Innovation Program Trust Fund and the Korea–World Bank Partnership Facility Trust Fund.

The authors of introductory chapter 1 are Mark A. Dutz and Carlos Rodríguez-Castelán, with contributions from Rogelio Granguillhome Ochoa and Samantha Lach on linking DTs to poverty reduction, from César Calderón and Catalina Cantú on the effect of the digital economy on growth and poverty reduction, from Guido Porto on an analytical framework on the effect of DTs on poorer households' income earning choices, and from Xavier Cirera, Marcio Cruz, Leonardo Iacovone, and Jesica Torres on quantifying the effect of COVID-19 on the private sector.

The authors of chapter 2 are Carlos Rodríguez-Castelán and Samantha Lach, with contributions from Rogelio Granguillhome Ochoa and Takaaki Masaki on determinants of internet adoption by households and individuals (based on the Enquête de Suivi de la Pauvreté au Sénégal and the 2018–19 Enquête Harmonisée sur les Conditions de Vie des Ménages—EHCVM); from Izak Atiyas and Toker Doganoglu on determinants of internet adoption by households (based on Research ICT Africa data); from Rogelio Granguillhome Ochoa and Takaaki Masaki on the analysis of the welfare effects of internet access; from Ted Enamorado, Takaaki Masaki, and Hernan Winkler on the effects of digital technology access on territorial inequalities; from Rogelio Granguillhome Ochoa on the ex-ante analysis of the welfare effects of competition in the telecommunication sector; and from Ed Oughton on policy options for affordable internet expansion (infrastructure sharing and regional integration).

The authors of chapter 3 are Marcio Cruz and Mark A. Dutz, with contributions from Xavier Cirera, Diego Comin, and Kyung Min Lee on the analysis of the Firm-Level Adoption of Technology survey; from Izak Atiyas on the determinants of DT adoption by micro enterprises; and from Jesica Torres and Trang Tran on entrepreneurship ecosystems. Additional contributions include the following: from Laurent Corthay on policies to improve regulations and access to markets (including a case study on CommAgri and Commango by Dalberg Group), from Mariana Vijil on diagnostics and policies to improve access to markets (focusing on supply chain reliability and the trade single window), from Laurent Gonnet on diagnostics and policies to improve access to financing (including a case study on EcobankPay), and from Alexandre Henry on mapping intermediary organizations supporting entrepreneurship.

This project was conducted under the overall guidance of Nathan Belete (Country Director) and benefited from useful guidance from Albert Zeufack (Chief Economist, Africa Region) and peer-review comments from Mary Hallward-Driemeier, Sandeep Mahajan, Carlo Maria Rossotto, and Federica Saliola. This project benefited from the advice of Najy Benhassine, Andrew Dabalen, Ejaz Ghani, Luc Lecuit, Johan Mistiaen, Sebastian-A Molineus, Lars Christian Moller, Consolate Rusagara, and Paolo Zacchia. The team also benefited from discussions with Louise Cord, former Country Director of Senegal, who initiated this task, and with Victor Ndiaye, CEO of Performances Group. The team thanks them for their helpful suggestions. A particularly strong thanks and recognition are owed to Louise Cord for having championed this task and supported it strongly through its conceptualization and initial phases of implementation. The team would like to thank other colleagues for contributions to this project, including Eva Ebion, Arthur Foch, Dieynaba Kane, Silvia Muzi, Djibril Ndoye, Sadibou Sylla, and Ibrahima Tall. Roselyne Mabudu provided outstanding administrative assistance. The team would also like to thank Mousa Blimpo, Tania Begazo Gomez, Georges Houngbonon, Tim Kelly, Markus Kitzmuller, Julio Loayza, and Cedric Okou for comments and suggestions on different components of this project. The team is grateful to Alison Gillwald and Onkokame Mothobi for making available the Research ICT Africa (RIA) ICT Access data from surveys of individuals/households and businesses for 2017–18 and for their support in the use of these data.

This project was a collaboration with the government of Senegal, with the Ministry of Economy, Planning and International Cooperation (MEPC); the Ministry of the Digital Economy and Telecommunications (MENT); the Ministry of Commerce and SMEs (MCPME); the Small and Medium Enterprise Development and Management Agency (ADEPME); the General-Delegation for Rapid Entrepreneurship (DER/FJ); the National Agency of Statistics and Demographics (ANSD); the Agency for Promotion of Investments and Strategic Works (APIX); and the General-Direction of Customs (DGD).

The team thanks in particular Pierre Ndiaye, Secretary General MEPC; Anta Ndoye, Technical Adviser, MEPC; M. Bamba Diop, Director General of the General Direction for Planning and Economic Policies, MEPC; Ndeye Maguatte Diouf, Director for Private Sector Development, MEPC; Idrissa Diabira, Director General, ADEPME; Mariame Kane, Head of Investment, DER/FJ; Thierno Sakho, Lead for Digital Economy, DER/FJ; Mamadou Lamine Ba, Director for

Investment Climate Reforms, APIX; Soda Diop, Head of Market, Agribusiness, APIX; Alioune Dione, Director of the IT Department of the Customs Agency; Modou Ngom, Director of Telecommunications, MENT; as well as Babacar Ndir, General Director; Mbaye Faye, Director of Economic Statistics and National Account; Alpha Wade, Head of the Partnership Management Unit; and Insa Sadio, Head of the Bureau of Business Statistics, all from ANSD.

Executive Summary

This book explores the extent of adoption and use of digital and complementary technologies in Senegal and describes pathways to increase their beneficial effects on economic transformation for jobs and poverty reduction. The main message of this book is that broader use of productivity-enhancing technologies by households and enterprises can generate better and more jobs, including for lower-skilled people, and support both the short-term objective of economic recovery and the government's long-term vision of economic transformation with more inclusive growth. But this result is not automatic. For this benefit to happen, Senegal should consider strengthening its policies—through targeted incentives to promote adoption and use by firms as well as to narrow deepening divides across enterprises and across households, in addition to ensuring availability of affordable infrastructure. This book leverages a novel survey that measures technology adoption at the firm level and a wide range of in-depth background studies.

Senegal needs better and more jobs for its growing population. More than 300,000 new jobs are required per year, projected to rise to 500,000 per year by 2050. To resume its trajectory of economic growth and make it more inclusive, Senegal needs to create better jobs for more people, the vast majority of whom are engaged in the informal sector. Indeed, less than 5 percent of the economically active population is in the formal private sector, and formal firms with 5 or more employees account for a mere 1 percent of all firms. Better and more jobs require better and more firms, including more productive and growing formal and informal enterprises.

Firms in Senegal display low levels of technology adoption, even though some capable informal enterprises show potential to jump the quality hurdle. The firm-level adoption of technology survey and analysis in this book exposes a large average technological gap between firms in Senegal and firms in Brazil (its state of Ceará), in the range of 36 percent and 30 percent for extensive (whether firms use it at all) and intensive (the most frequently applied) uses, respectively, of better technologies such as for business administration and marketing. Except for a small share of firms, Senegalese enterprises still mostly use manual procedures and predigital technologies to perform general or sector-specific business functions. Micro informal enterprises lag even further. Although 30 percent of larger firms use smartphones, only 18 percent of micro informal enterprises do so.

Only about 6 percent use management software tools. However, more than 27 percent of young-women-owned micro firms use a smartphone, more than 12 percent use inventory control/point-of-sale (POS) software, and more than 24 percent use the internet to better understand their customers for marketing and sales—the highest shares respectively for each of these digital technologies across age and gender dimensions. This finding highlights a sizable potential for some capable informal enterprises as well as formal firms and fast-growing start-ups, including young-women-owned businesses, to upgrade technologies, strengthen their capabilities, and join the modern economy.

The benefits from technology adoption are significant. Digital technologies are an enabler of economywide productivity and job growth by catalyzing adoption of complementary technologies, including many not accessible without digital infrastructure. For households, mobile internet coverage is associated with 14 percent higher total consumption and 26 percent higher nonfood consumption, as well as a 10 percent lower extreme poverty rate—and jobs with higher earnings. Firms with better technologies have higher levels of productivity, generate more jobs, and increase the share of unskilled workers on their payroll, on average: an increase in technological sophistication for business functions that the firm uses most intensively, such as using standard Excel spreadsheet software rather than writing by hand for accounting and inventory control, is associated with a 14 percent increase in the number of workers in the average firm. For micro informal firms, using a smartphone per se does not generate more jobs. What matters jointly for productivity, sales, and jobs are digital technologies for management functions such as inventory control/POS software—in addition to having electricity and a loan. Digital technologies also facilitate the delivery of a wide range of public services.

Supporting economic recovery through more inclusive and productive job creation requires government policies to improve the availability of affordable digital infrastructure on the supply side, and bolster the technology uptake of firms by strengthening their capabilities as well as their access to markets and finance on the demand side. Enterprise response to COVID-19 so far is deepening digital divides between larger and smaller firms, with 36 percent of large firms reporting having invested in new digital solutions in the initial months while only 8 percent of small firms have done so. If such digital divides remain unaddressed, these diverging investment trends are expected to widen productivity, sales, and owner and worker income gaps over time. To enhance the availability of affordable broadband, Senegal needs to ensure affordable access to electricity, promote universal coverage of 3G/4G supported by increased competition in the telecommunications market, ensure effective infrastructure sharing, and leverage private investments, among other measures. To boost productive use of digital technologies by enterprises, Senegal should institutionalize technology upgrading and management/worker capability support programs. Customizations of such programs according to enterprise capabilities and needs are warranted for formal start-ups and larger firms with the potential to become globally competitive national champions, and for those micro-size informal enterprises willing and able to learn and adopt technologies to attain consistently higher quality levels—thereby providing significantly higher benefits from eventual formalization. To yield results, these programs should be anchored in industry value chains, such as specific horticulture products, and supported by a punctual execution plan and an effective delivery unit. Downstream wholesale/retail and exporting buyers and digital tool providers will help pull sustainable quality upgrading through the value chain. Additional business environment reforms on which this book elaborates are necessary to strengthen the entrepreneurship support ecosystem and boost firms' access to finance. Finally, greater adoption and use by households is essential to expand the productive capacity of workers and micro-entrepreneurs and connect them to better jobs, as well as to increase purchases by households of enterprise offerings, many of which cannot be accessed without smartphones. Demand-side policies to stimulate more inclusive use include reducing budget constraints and encouraging zero pricing for messaging apps for poor households, as well as incentivizing the creation and uptake of easy-to-use apps that respond to the needs of lower-income, lower-skilled workers and other vulnerable and excluded groups.

Abbreviations

ADEPME	Agency for Development and Supervision of Small and Medium Enterprises (Agence de Développement et d'Encadrement des Petites et Moyennes Entreprises)
ADIE	State Informatics Agency (Agence de l'Informatique de l'État)
ADSL	asymmetric digital subscriber line
AI	artificial intelligence
ANSD	National Agency of Statistics and Demographics (Agence Nationale de la Statistique et de la Démographie)
APIX	Agency for Promotion of Investments and Strategic Works
ARTP	Regulatory Authority for Telecommunications and Post (Autorité de Régulation des Télécommunications et des Postes)
B2G	business-to-government
BCEAO	Central Bank of West African States (Banque Centrale des États de l'Afrique de l'Ouest)
BMN	Business Upgrading Agency (Bureau de Mise à Niveau des Entreprises)
CDF	cumulative distribution function
COV-BPS	COVID-19 Business Pulse Survey
CRM	Customer Relationship Management
D4Ag	Digitization for Agriculture
DE4A	Digital Economy for Africa
DER	Delegation for Rapid Entrepreneurship
DT	digital technology
ECOWAS	Economic Community of West African States
EHCVM 2018/2019	Enquête Harmonisée sur les Conditions de Vie des Ménages 2018/2019 survey
ELEPS	Light Experimental Poverty Assessment Survey
ERP	Enterprise Resource Planning
FAT	Firm-Level Adoption of Technology survey

FDSUT	Universal Services Fund (Fonds de Développement du Service Universel des Télécommunications)
FONGIP	Guarantee Fund for Priority Investment
G2B	government-to-business
GBF	general business function
GPT	general purpose technology
GSMA	Global System for Mobile Communications Association
GVC	global value chain
IBGE	Brazilian Institute of Geography and Statistics
ICT	information and communication technology
ID4D	Identification for Development initiative, World Bank
IoT	internet of things
ISP	internet service provider
ITU	International Telecommunication Union
KPI	key performance indicator
M&E	monitoring and evaluation
MEs	mesas ejecutivas (public-private working groups in Peru)
MFI	microfinance institution
MNE	multinational enterprise
MSME	micro, small, and medium enterprises
NAEMA	Nomenclature d'activités des états membres d'AFRI-STAT, Observatoire Économique et Statistique
OECD	Organisation for Economic Co-operation and Development
PAP2A	Plan d'Actions Prioritaires II Ajusté et Accéléré
PAP2/PSE	Plan d'Actions Prioritaires II du Plan Sénégal Émergent
PED	price elasticity of demand
PER	Public Expenditure Review
POS	point-of-sale (software)
PPP	public-private partnership
PREAC3	Programme de Réforme de l'Environnement des Affaires et de la Compétitivité (phase 3)
PSE	Plan Sénégal Émergent
RGE	Recensement Général des Entreprises
RIA	Research ICT Africa
SENELEC	Senegal National Electricity Agency
SGPR	Secretary-General of the Presidency
SMEs	small and medium enterprises
SSA	Sub-Saharan Africa
SSBF	sector-specific business function
SN2025	Sénégal Numérique 2016–2025
SYSCOA	West African Accounting System
VAT	value-added tax
WAEMU	West African Economic and Monetary Union
WELCOM	Welfare and Competition simulation tool
WTO	World Trade Organization

Overview and Main Recommendations

INTRODUCTION

Senegal needs better and more jobs for its growing population. Stagnant productivity levels and persistent inequalities, combined with the COVID-19 crisis, threaten Senegal's aspiration to become an upper-middle-income country with greater inclusion by 2035. In view of Senegal's growing labor supply, its main job challenge is to sustainably reduce poverty through the creation of more than 300,000 jobs each year. Absent demographic changes, this number is projected to rise to 500,000 per year by 2050.

To ensure that accelerated job creation is productive and inclusive, policies are needed to boost use of better technologies across firms and to address deepening digital and other divides across households and across enterprises. The COVID-19 crisis presents an opportunity to "build back better." Evidence is emerging that the use of digital technologies (DTs)—defined broadly to include not only smartphones and the internet but also a variety of more specialized productivity-enhancing digital solutions (World Bank 2016)—can be an entry point to enable economies to better respond to emergencies as well as to grow. Digital sources of growth largely come from cost reduction, efficiencies, and enhanced capabilities associated with households and enterprises using better technologies, with many of the latest analog technologies now also incorporating links to the internet and better data use. To tap into these sources of growth, Senegal needs a mix of supply-side policies to support affordable availability of DTs and complementary technologies, as well as demand-side incentives and capability-boosting programs to stimulate their adoption and productive use.

Building on government priorities, this book presents a diagnostic of current trends and drivers of DT adoption and use in Senegal and explores how the country can leverage DTs to boost economic transformation and jobs, while adequately mitigating associated risks. The book's main objective is to provide new data and analyses to support the efforts of the government of Senegal to spur inclusive growth through the adoption of appropriate updated technologies, while avoiding deepening digital divides between geographic zones, enterprises, and groups of people. In particular, this book aims to help address the first challenge identified in the adjusted and accelerated Plan d'Actions Prioritaires II du

Plan Sénégal Émergent (PAP2/PSE), namely, the development of a competitive, inclusive, and resilient economy.[1] This book complements the PAP2/PSE's analysis,[2] which identifies productivity as the key to structural transformation, noting that Senegal is still lagging in relation to comparator countries.[3] The book is also aligned with the orientations of the Sénégal Numérique 2016–25 (SN2025) strategy, which views the industry that generates, produces, and diffuses DTs as a critical sector in Senegal's economic and social development.[4] DTs are an enabler of economy-wide productivity gains and job growth by catalyzing the adoption of complementary technologies, including many not accessible without digital infrastructure. For higher-quality agriculture, for example, taking advantage of the IoT (internet of things) requires investments in DTs but also in "things" such as tractors and irrigation systems equipped with sensors, smartphones to access weather forecasts and upload pictures of unusual plant diseases, and appropriate apps with video to enable even illiterate farmers to integrate into formal value chains, learn from downstream buyers and upstream seed providers, and have better access to financing and markets.

This book seeks to answer three main policy questions:

1. What is the extent of digital technology adoption and what are the main barriers preventing a broader adoption by households and enterprises?

2. What are the effects of digital technology adoption on household welfare and enterprise productivity?

3. What are key areas for policy intervention that could promote greater adoption and intensity of use of digital technologies?

While answering these questions, the book uses a conceptual framework based on five channels leading from affordable DT availability to faster and more inclusive growth with positive welfare effects. The pathways through which DTs can enhance the inclusive income generation capacity of households and enterprises are labor incomes, capital incomes, consumer surplus, the tax-transfer system, and nonmonetary factors (see figure O.1). Of these, the book focuses on three channels: (a) labor incomes (DTs enabling better jobs for more people, including lower-income people); (b) capital incomes (including profits earned by entrepreneurs and owners of larger firms as well as self-employed subsistence farmers and owners of other smaller formal and informal enterprises); and (c) improved consumption opportunities (through lower prices and a higher quality and greater variety of products). The report also refers to how better access to financing interacts with these channels to improve efficiency and equity outcomes. The book does not focus on how DTs facilitate the delivery of a wide range of public services, neither the effect of DT upgrading on e-government through the income transfer system nor the effect of e-government and other DT improvements on nonmonetary gains.

The overall message of the book is that broader adoption of better technologies by households and enterprises can produce better jobs for more people and support the long-term objective of economic transformation for more inclusive growth. None of this economic improvement is automatic. Greater digital inclusion to connect people to markets requires more efforts to broaden affordable and high-quality internet access and use by all, without which none of the other more sophisticated technologies can be accessed. Some of the measures to achieve digital inclusion include increasing

FIGURE O.1
Conceptual framework: From DT availability to inclusive growth

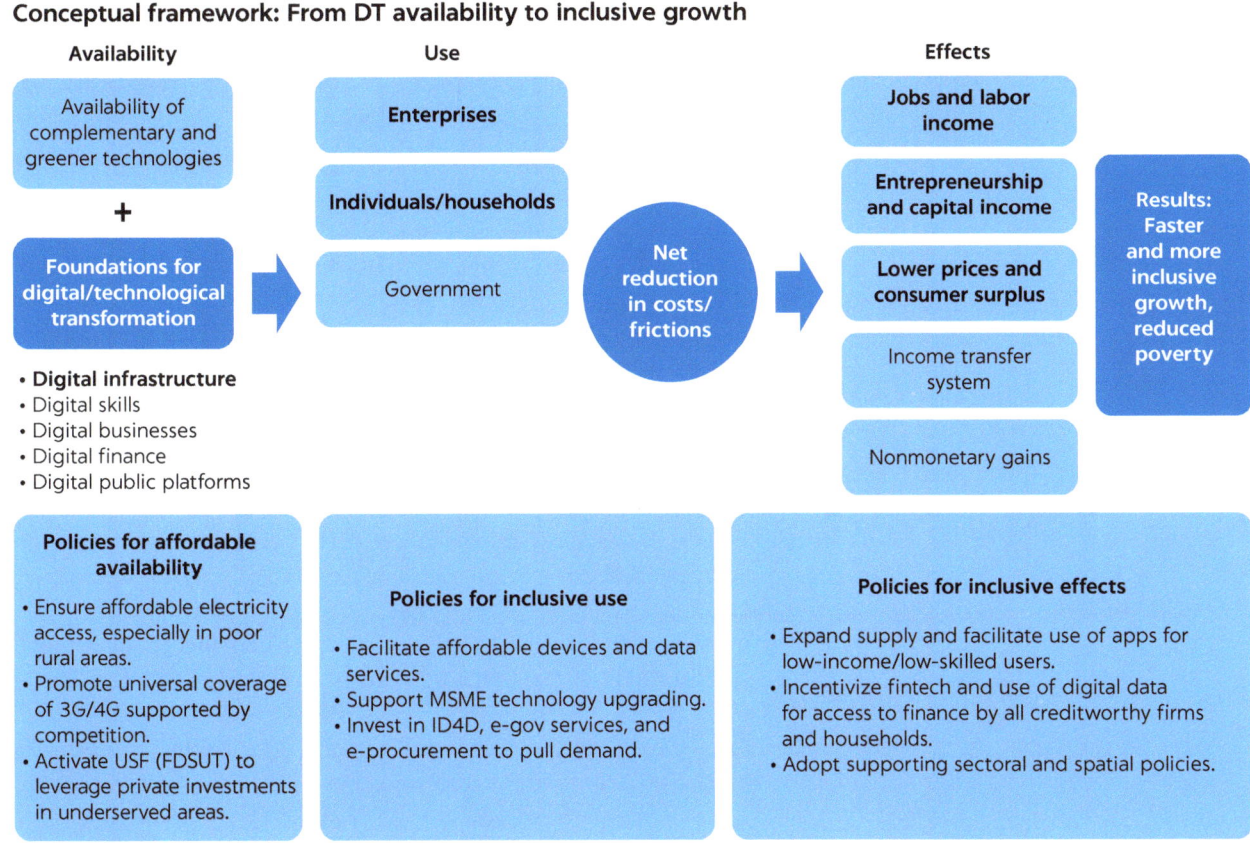

Source: World Bank.
Note: Text in bold represents the focus areas of this book. Also see figure 1.7.

competition in the telecommunications sector and promoting affordable access and use of DTs by excluded populations through useful apps and services in local languages and videos for illiterate people. Gaining *better jobs* for more people requires support for enterprises' use of better technologies, supplemented by a more conducive environment for start-ups with emphasis on access to finance. Generating *more jobs* through the adoption of productivity-enhancing technologies, including for lower-skilled people, requires—in addition to more new firms—larger volumes of production by existing firms that should be forthcoming from cost reductions and quality improvements enabled by these technologies. These advances, in turn, require the following: competition in input and output markets for more efficient firms to expand, sufficient responsiveness of consumer demand to the lower prices stimulated by technologies and product competition, and adopted technologies to be sufficiently complementary with lower-skilled workers so that their jobs are not eliminated but rather transformed into new tasks for which they can build the needed capabilities as they work. Greater digital inclusion and accelerated growth require going beyond "good policies" that merely cushion the blows of an economic shock and then continue with business as usual. They require "great policies" anchored in deep structural reforms that support technological transformation and skill upgrading, sectoral reallocation, and spatial integration of assets across the country and the region toward more productive, job-enhancing uses, including for low-skilled, low-income workers.

This overall message becomes even more important and urgent in the context of the COVID recovery. A recent set of Business Pulse Surveys conducted by the World Bank in Senegal suggests a significant adverse economic impact from the pandemic. At the time of the first survey in early May 2020, 27 percent of businesses were only partially open and 15 percent were temporarily closed because of containment measures. More than 40 percent of workers in these businesses were facing high levels of vulnerability. The negative effect on sales has been large and widespread. Almost all (90 percent) of businesses experienced a decline in sales. Small firms were disproportionately affected by the shock, with an estimated drop of 55 percent in sales on average over the 30 days before the survey, relative to the same period in 2019. At the time of the second survey in December 2020–January 2021, still more than three-quarters of all firms (76 percent) continued to experience a decline in sales. In response to the shock, and until January 2021, almost one-half (40 percent) of firms across all size groupings started to use or increased the use of DTs for business purposes. At the same time, although the pandemic has accelerated DT use and made DTs more indispensable, it also has highlighted persistent and growing digital divides: the share of medium and large firms investing in DT solutions in response to COVID, at 33 and 36 percent of these size groups, is more than twice as large as small firms, at 14 percent (figure O.2). On the one hand, the rise in demand for DTs by households and enterprises following the onset of COVID could help with the success of public policies that promote technological upgrading for better jobs for more people. On the other hand, if these digital divides remain unaddressed and greater supply-side affordability and more demand-side use incentives are not forthcoming, these diverging investment trends are expected to widen productivity, sales, and owner and worker income gaps over time.

FIGURE O.2

COVID-19 has accelerated use of DTs but is increasing digital divides

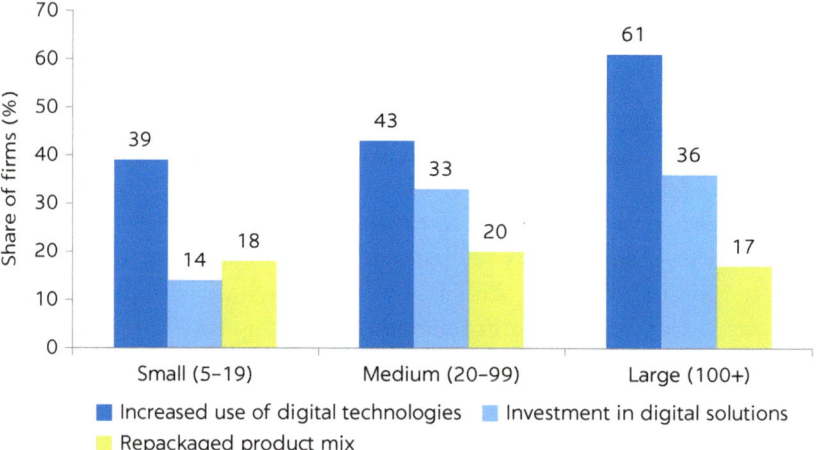

Source: COV-BPS Senegal (2021).
Note: See figure 1.10, panel b.

OVERVIEW OF CHAPTER 2—HOUSEHOLDS: WELFARE EFFECTS OF DIGITAL TECHNOLOGIES

To take advantage of digital technologies, households first need to adopt them. Uptake is influenced by a range of factors both at the individual level—including income, education, area of residence, and other socioeconomic characteristics—and at the country level, for example, coverage and prices of DTs.

During the past decade, Senegal has achieved progress in expanding broadband coverage; however, the affordability, use, and quality of services remain relatively low. Whereas Sub-Saharan Africa is the region with the lowest internet penetration in the world (18.7 percent), Senegal's penetration, at 29.6 percent, is relatively high; yet it remains behind the global average of 49 percent.[5] According to the Global System for Mobile Communications Association (GSMA), Senegal's unique mobile internet connections rate, at 31 percent of the population in 2019, lags behind regional leaders, such as Ghana and South Africa, and is far behind high performers in other regions. The mobile broadband cost to households—at 3.1 percent of monthly income in 2019 for a data-only package—lies slightly above the levels established by the Broadband Commission for Sustainable Development at less than 2 percent of monthly income. The quality of service is low, while both broadband density and internet bandwidth lag behind regional comparators such as Côte d'Ivoire, Ghana, and Nigeria.

At the household and individual levels, the main factors determining mobile internet adoption include household income (measured by consumption per capita), price of mobile internet, age, gender, tertiary education, language, living in an urban area, employment sector, asset ownership, and access to electricity. Increasing monthly per capita expenditures on average by CFAF 40,000 (about US$72)[6] per year (equivalent to 1 standard deviation) would increase mobile internet adoption by 9.3 percentage points. The affordability of mobile internet services also plays a role: an average decline in the monthly price of mobile internet of CFAF 1,100 (US$2.00) (equivalent to 1 standard deviation) would increase adoption by 2.0 percentage points. The evidence also points to gaps across socioeconomic groups. Being a woman lowers the likelihood of adoption by 6 percentage points, while having tertiary education or higher increases this probability by 16 percentage points. Individuals ages 25–40 years are 21 percentage points more likely to have access to mobile internet—and so are those living in urban areas (by 4.0 percentage points), reflecting the existence of a rural/urban divide. These findings highlight the need for gender, skill, age, and location-specific approaches to promote DT adoption. Language also matters, because individuals who can read and write in French are 13 percentage points more likely to access the internet through their mobile phones. In addition, complementary analysis using RIA data finds that online social networks appear to be an important driver of uptake: an increase from one to five in the number of friends who use messaging applications is associated with a rise in the adoption probability from 2.5 percent to 37 percent.

Once households or individuals adopt DTs, these technologies can have important effects on welfare. As shown in figure O.3, mobile internet coverage in Senegal is associated with a 14 percent higher total consumption (and about 26 percent higher nonfood consumption) for covered households, as well as with a 10 percent lower extreme poverty rate. These results are in line with those of a recent study based on more detailed panel data from Nigeria, which finds a 9 percent increase in consumption and a nearly 7 percentage point

FIGURE O.3

Mobile internet coverage is associated with higher consumption and lower poverty

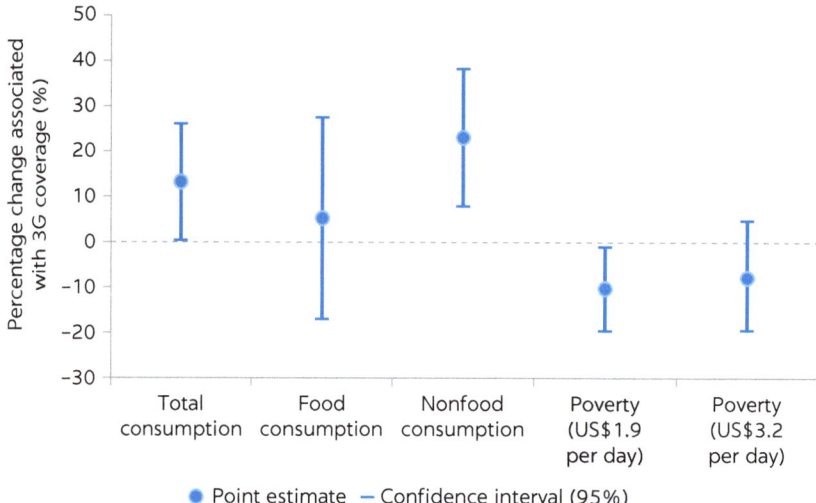

Source: Masaki, Granguillhome Ochoa, and Rodríguez-Castelán 2020.
Note: See figure 2.3.

decrease in poverty as a causal result of the internet after three years of coverage (Bahia et al. 2020). The analysis for Senegal also finds that welfare effects differ across groups, with greater magnitudes in urban areas and for young and male-headed households.

One of the main channels for mobile broadband internet to translate into welfare improvements is the labor market. On the basis of household surveys and internet coverage data, this book provides evidence that mobile broadband coverage is positively correlated with wage and salaried employment, formal employment, and monthly earnings. This finding has important implications for the potential role of DTs in contributing to improved labor market outcomes and job creation. There is also evidence that, at the local level, expansion in broadband coverage is associated with modest poverty reduction and improvement in local economic activity. Although these analyses are not without limitations, they provide useful insights to inform policy discussions on strategies to enhance the welfare effects of mobile broadband internet access and reduce the digital divide across territories and socioeconomic groups.

The government should focus on improving the availability of affordable digital infrastructure, underpinned by improved access to electricity, particularly in rural areas, not just to prevent widening digital divides but also to actively narrow them. Policies geared toward universal and affordable broadband and electricity coverage are among the most consequential to reach excluded populations, including people living in remote and rural areas. In this context, accelerating the actions to activate Senegal's Universal Services Fund (Fonds de Développement du Service Universel des Télécommunications, FDSUT) and delegating the management of the publicly owned fiber-optic infrastructure to a private sector wholesale operator could help reduce the digital divide by leveraging private investments. Deepening ongoing reforms to increase competition in digital infrastructure and service provision, including by stimulating market entry of new service providers, while reducing the market power of dominant operators, can also help reduce tariffs for all. Importantly, a recently adopted infrastructure sharing policy for rural areas should be implemented with adequate safeguards to prevent anticompetitive conduct. To take advantage of sizable regional benefits by spurring the development of fully regional data markets, Senegal should also consider playing a leadership role in increasing regional harmonization, including ensuring coordination at the supranational level regarding the West African Economic and Monetary Union (WAEMU) electronic communications framework and Economic Community of West African States (ECOWAS) rules, implementing fully the ECOWAS roaming regulations, and advocating for the elimination of restrictive (or lack of) rules across ECOWAS countries.

Concomitant demand-side policies are necessary to enhance the positive effects of mobile broadband internet access and ensure its productive use. Policies that reduce budget constraints for households, such as social assistance transfers and the removal of barriers to financial inclusion, are critical to improving mobile internet uptake. Encouraging zero pricing for messaging applications—for instance, through a low-usage package attractive only to low-income households—can help take advantage of network effects. Promoting local language content and continuing to invest in digital skills can ensure that mobile coverage translates into effective use. Importantly, vulnerable groups may require specific attention, including approaches that are tailored to individuals' gender, age, and location.

Although a detailed regulatory analysis goes beyond the scope of this book, some regulatory aspects on both the supply and the demand side are touched upon in the analysis. The book thereby highlights the importance of regulatory frameworks for expanded adoption and use of DTs. Particularly critical are promoting market entry and innovation and adopting a broader approach to data, digital business models, and privacy/data sharing concerns, which help shape the inclusiveness of digitally driven growth. An in-depth analysis of these issues is provided by the *World Development Report 2021: Data for Better Lives*.

OVERVIEW OF CHAPTER 3—ENTERPRISES: INNOVATION FOR BETTER JOBS FOR MORE PEOPLE

Better jobs for more people require better and more firms. Better and more firms, in turn, require the following: technology upgrading and use of new technologies by existing enterprises, more new firms and entrepreneurship, and better access to financing. Senegal needs to focus its support policies on both formal and informal firms.[7] Formal private sector jobs account for only 5 percent of the active working-age population, or fewer than 200,000 jobs. Meanwhile, the working-age population is increasing by 300,000 per year and is expected to rise to roughly 430,000 per year by 2030. So, to support more inclusive growth, the policy focus also needs to include boosting productivity of informal firms, with public support for them sufficiently attractive to lead those firms to eventual formalization.

The following overview describes the three essential ways to support better and more firms.

Technology upgrading support for better firms

Senegalese firms lag in the adoption of more sophisticated technologies. Compared with the state of Ceará in Brazil,[8] Senegal has a gap of 36 percent in adopting and using any type of more sophisticated technologies for general business functions (for example, keeping accounting records by hand versus by a simple Excel spreadsheet versus using a more sophisticated Enterprise Resource Planning [ERP] solution), and a gap of 30 percent for the most frequently used technology. Senegal's average firm also lags behind Vietnam in technological sophistication in these general business functions, both on the extensive margin (whether the firm uses the technology at all) and on the intensive margin (whether it is the technology used most intensively by the firm). Moreover, Senegal lags in the adoption and use of more sophisticated technologies for industry-specific business functions. As an example, land preparation for farming remains largely manually done, with only a small share of firms using a single-axle tractor, and almost none using more sophisticated versions of GPS-guided tractors with internet-of-things sensors.

Adopting better technologies is associated with faster job growth. Almost all more sophisticated technologies today have some digital elements. Firms that use more sophisticated technologies are more productive than those that use less sophisticated ones. Importantly, firms that use more sophisticated technologies also have faster job growth than those that use less sophisticated technologies. Interestingly, firms that use more sophisticated technologies intensively for general business functions, such as for business administration and production planning, have faster job growth (a 1-point increase in the technology adoption index is associated with a 14 percent increase in the number of workers in the average firm) than those that use them less frequently, controlling for initial size, age, sector, region, exporting status, and foreign or domestic ownership (see figure O.4). As well, the uses of internal-to-the-firm DTs for business administration and production planning have an association with higher average job growth than the uses of external-to-the-firm DTs for upstream sourcing and downstream

FIGURE O.4

Firms that use more sophisticated digital technologies have faster job growth

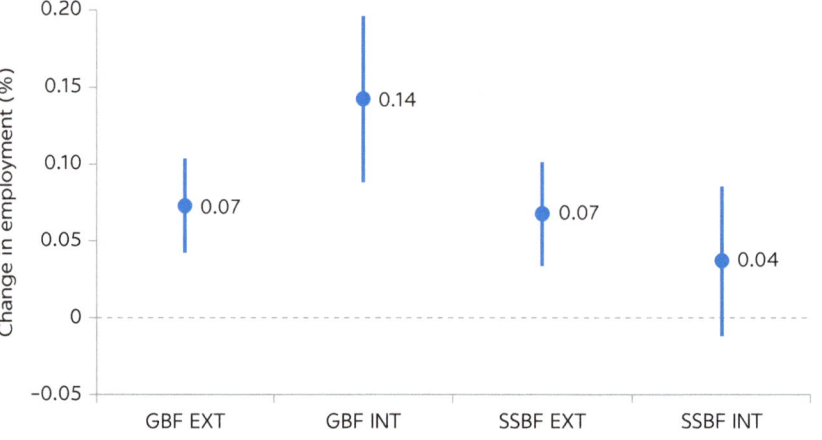

Source: Cirera et al. 2021.
Note: See figure 3.14. EXT= extensive margin; GBF = general business function; INT = intensive margin; SSBF = sector-specific business function.

marketing, sales, and payment methods. This fact could be partly explained by the latter requiring a more widespread and effective ecosystem of adoption, including other upstream and downstream firms and individuals as users. And critically for reducing digital divides, job growth from technology upgrading is inclusive: firms with more sophisticated levels of technologies not only generate more jobs but also increase the share of unskilled workers on their payroll.

Micro informal enterprises lag even further in technology adoption.[9] This fact, conversely, means that there may be greater potential for technology upgrading and continual learning, productivity, and sales as well as job increases for those informal firms that are able to jump the quality hurdle. Though 30 percent of larger firms (including informal ones) use smartphones, only 18 percent of micro informal enterprises do so. Fewer than 6 percent of these firms ever had a loan, and about the same share use general business function management software tools instead of writing account information and so forth on paper and not being able to consider what it means for company profits and growth.[10] Those informal enterprises that use more specialized DTs, compared with those that do not use them, have higher average levels of labor productivity and total sales, and are more likely to export products. They also generate higher average numbers of jobs (have a larger firm size) and earn higher per-owner average incomes. Importantly, the use of inventory control/POS software as a proxy for internal-to-the-firm better management practices is the only consistently significant conditional correlate of productivity, sales, and jobs—in addition to having electricity and having a loan. This observation suggests that the adoption of relatively simple DTs to improve basic management functions could be an important component of a more inclusive job growth agenda.

The main barriers to adoption of better technologies are weak access to finance, low levels of firm capabilities, and insufficient access to markets with robust competition. Lack of finance is most pronounced for smaller firms and is most associated with firms not investing more intensively in technology upgrading. The next important barrier is the lack of internal capabilities to get information (on what to buy) and knowledge (on how to use the purchased technologies). That barrier is followed by uncertainty of demand, such as whether input and output markets, including public procurement, can be accessed by smaller firms to enable expansion. It is also linked to upstream and downstream market competition bottlenecks, such as a nonlevel playing field with favored access for some players. Most firms, especially smaller and informal ones, do not benefit from consulting advice. Informal firms believe they are better than they are (they don't know what they don't know). These barriers also align with the barriers faced by micro informal enterprises—having a loan is the largest significant conditional correlate of smartphone adoption, followed by vocational training of business owners and access to electricity.

The government should focus on institutionalizing technology extension and management/worker capability support programs, and on continually improving these programs through experimentation and learning. These programs should be customized to address the different needs of more technologically sophisticated firms that are closer to the frontier and of less technologically sophisticated micro-size and small informal farms and firms able to learn and jump the quality hurdle to join the modern economy. A more technologically sophisticated program should support firms with more sophisticated capabilities and needs in the highest-potential ecosystems to benefit from agglomeration and network economies and spatial spillovers. Priority ecosystems include complementary value chains in agribusiness, wholesale/retail trade, and digital services: average sales per worker in high-potential agribusiness ecosystems are 5 times higher than in maturing ecosystems. A customized part of the program should be offered to firms with less sophisticated capabilities, including informal firms that through their actions indicate a willingness and an ability to improve productivity and formalize over time as they benefit from sufficiently attractive productivity support programs. For example, participation by informal farms in cooperatives receiving public support to upgrade quality standards could be made contingent on meeting minimum quality and productivity requirements and demonstrating an ability to grow and to eventually formalize. Businesses could benefit from current programs,[11] appropriately modified and anchored in industry value chains such as specific horticulture products, focused on solving coordination problems and facilitating technology transfer, learning, and market access from larger formal downstream buyers and upstream sellers as well as between smaller firms. Such programs need to be supported by a punctual execution plan and an effective delivery unit. The design and scaled use of DTs should be supported, not only for value-chain-specific information, financing, and management tools but also to reduce implementation costs in the provision of consulting and monitoring services.

Start-up entrepreneurship support for more firms

Senegal lags behind comparators in the dynamism of new firm creation. According to the 2018 World Bank Entrepreneurship Database, the share of new registered businesses per thousand working-age population in Senegal is 0.5 versus more than twice as many in Kenya (1.1) and Côte d'Ivoire (1.5). Increasing the number and quality of new firms can have a significant influence on job creation over time. Senegal also lags behind comparator countries—such as Côte d'Ivoire, Kenya, and Morocco—in total factor productivity, as highlighted in the PAP2/PSE.[12] The main barriers faced by productive start-up entrepreneurs in high-potential ecosystems are insufficient access to markets, difficulty in distributing and selling final products, and burdensome regulations. Weak capabilities are also a constraint, associated with the lack of access to technology and low levels of human capital.

Some foundational legal and regulatory frameworks need updating to stimulate the creation of better and more firms. To help spur entrepreneurial dynamism, the government should focus on reforms that facilitate ease of entry and growth as well as ease of exit and reentry if the initial investments and business models are unsuccessful. As highlighted in the Programme de Réforme de l'Environnement des Affaires et de la Compétitivité now entering its third phase (PREAC3), deepening the digitization of G2B transactions is a priority—though it also requires ensuring affordable internet availability to all firms wanting to avail themselves of these opportunities.[13] This effort is particularly important for the payment of taxes and fees. That area has registered substantial progress in recent years but still constitutes one of the biggest constraints to business growth and formalization. Improving trade facilitation and logistics—aided by DT solutions for strengthened risk management, centralized logistics cost payments, tracking and tracing, and scaling up of the vehicle booking system in the port of Dakar—could support imports of key products, including those critical for the COVID-19 response. It could also improve livelihoods in the longer term by facilitating participation in global value chains, business investments, export performance, and job growth. Finally, the digitization of G2B transactions should include a transparent and fair public procurement platform, enabling entrepreneurs, especially smaller firms, to participate in tenders. Additional priorities include labor regulations, which have proven burdensome for many formal businesses and fail to effectively support formal employment creation. Access to land is another area where deeper reforms should continue to be pursued, in particular by streamlining allocation procedures of land titles and ensuring transparent and reliable land rights inventory, secured land rights deliberations, and determination of the rental value of built properties. Strengthening the regulatory framework governing competition at the national level is another priority, given its current de facto lack of effectiveness, with the need to clarify mandates with the WAEMU regional framework.[14]

Beyond economywide reforms, policy attention should be focused on addressing key barriers that are specific to high-potential ecosystem value chains. These barriers include human capital in agriculture and access to technology and finance for the digital economy. Many resources have already been allocated to supporting start-ups and digital solutions. The allocation of future resources would benefit from a screening on the basis of business priorities in high-potential ecosystems. Solutions need to be designed to solve problems in specific value chains, for example, the mango value chain.[15]

Financing support for better and more firms

Technology upgrading and start-up entrepreneurship need to be enabled by financing. Senegal has a gap of US$1 billion in access to finance for micro, small, and medium enterprises (MSMEs), with needs for both debt and equity finance. Across WAEMU, firms in Senegal report being the most financially constrained. The main barriers are lack of competition in the banking sector, lack of credit infrastructure, a poor legal framework, and insufficient public interventions. Moreover, Senegal has neither a partial credit guarantee scheme nor a public–private venture capital fund to create a pipeline of start-ups. Opportunities to take advantage of e-wallets are still minimal, with only 7 percent of the adult population having received an e-payment from the government.[16] Overall, despite a few DT-based initiatives in finance, such as EcobankPay, financial DT solutions are not widespread in Senegal.[17]

Financing support could include a matching grant fund and financing of de-risking mechanisms. A matching grant fund could be used to support a range of initiatives, including (a) the digitization of microfinance institutions (MFIs) through a mutualized digital core banking system; (b) the expansion of rural mobile money agent networks; (c) the adoption of USSD codes by value-added service providers; (d) the digitization of business-to-government (B2G) payments such as the rural family support scheme; and (e) a package of measures supporting micro, small, and medium enterprises (MSME) financing, including an expansion of the small and medium enterprises (SMEs) rating system by ADEPME, the creation of a crowdfunding platform, and e-insurance. Financing of de-risking mechanisms could include support to the capitalization and operationalization of a start-up public-private risk capital fund, and a sustainable MSME guarantee fund. Fintech solutions, such as digital credit scoring services, should help overcome transaction costs and low capabilities (for example, less-prepared business plans) by allowing financiers to extend e-credit based on firms' credible business transaction records and other e-based criteria. Government support could also include the setting up of quicker out-of-court settlement procedures for firms that face solvency problems in a more unpredictable post-COVID-19 environment. See table O.1.

TABLE O.1 **Summary of main policy recommendations**

AFFORDABLE INTERNET AVAILABILITY	TIME HORIZON
Enhance access to affordable electricity, especially in poor rural areas.	Short term
Deepen ongoing reforms to boost competition in digital infrastructure and service provision as a way to reduce consumer prices and improve quality of services.	Short term
Delegate management of public fiber-optic network to private, wholesale infrastructure operator.	Short term
Implement the recently adopted infrastructure-sharing policy with adequate anticompetitive conduct safeguards.	Short term
Activate FDSUT to reduce digital divides by leveraging private investments.	Short term
Play a leadership role in increasing regional harmonization, including ECOWAS and WAEMU pro-competition and data market rules.	Medium term
HOUSEHOLDS: PRODUCTIVE USE OF DTs	
Reduce budget constraints for poor households, including through social assistance transfers and financial inclusion.	Short term
Encourage zero pricing for messaging apps for lowest-income households (and entrepreneurs) to take advantage of network effects.	Short term
Promote local language content and continue investing in education and basic digital skills.	Medium term
ENTERPRISES: PRODUCTIVE USE OF DTs	
1. Technology Upgrading and Support for Better Firms	
Institutionalize technology upgrading and management/worker capability support programs for firms, with a public-private dialogue mechanism and an effective delivery unit.	Short term
Build institutional mechanisms to continually improve business support programs through experimentation, evaluation, and learning, including by leveraging digital platforms and firm-level benchmarking based on existing microdata for Senegal and abroad.	Medium term
2. Start-Up Entrepreneurship Support for More Firms	
Deepen reforms to enhance competition in input and output markets to facilitate firms' entry and expansion as well as exit and reentry.	Short term
Deepen digitization of G2B transactions, including by improving access by all firms to tender for government procurement contracts.	Short term
Improve trade facilitation and logistics through DT solutions focused on risk management of controls on movement of goods across borders and centralized payment of all logistics costs.	Medium term
Promote mechanisms to facilitate networking and interaction within high-potential local ecosystems across different value chains and digital solution providers.	Medium term
3. Financing Support for Better and More Firms	
Strengthen a digital MSME scoring system, including for informal firms, to extend e-credits based on transaction records.	Short term
Create a new SME financing mechanism to support high-growth SMEs.	Short term
Increase effectiveness of the partial credit guarantee fund to facilitate sustainable MSME funding.	Short term
Support crowdfunding platforms and e-insurance products.	Short term

Source: World Bank.
Note: DT= digital technology; ECOWAS = Economic Community of West African States; FDSUT = Fonds de Développement du Service Universel des Télécommunications; G2B = government to business; MSME = micro, small, and medium enterprise; SME = small and medium enterprise; WAEMU = West African Economic and Monetary Union.

LOOKING FORWARD

The work summarized in this book could make important contributions to technical capacity building and inclusive technology policies in Senegal and SSA. The book team implemented the Firm-Level Adoption of Technology (FAT) survey in the country in close collaboration with the National Agency of

Statistics and Demographics (Agence Nationale de la Statistique et de la Démographie, ANSD). The FAT survey may be incorporated into the standard set of firm-level surveys implemented regularly by ANSD because of the value added of evidence-based policy making supported by this work. The complementary survey of largely informal, micro firms compiled by Research ICT Africa (RIA) in 2017–18 should be replicated in panel format, including a larger number of firms over time.[18] In addition, the work helped implement the COVID Business Pulse Survey (BPS), which has provided a rapid diagnostic of the effects of COVID-19 on firms and has identified characteristics allowing firms to cope during the crisis. Future rounds of the BPS would allow some of the same firms to be tracked over time, generating deeper insights about the connection between policies, firm behavior, and better jobs for more people. Moreover, the book presents a new methodology to identify local entrepreneurship ecosystems using firm-level census data, which is now being replicated in a few other countries, including Kenya. Finally, the work has contributed to the technical debate on the importance of DTs for achieving inclusive growth in Senegal by means of integrating different sources of data, including household survey data, firm-level data, customs data, and digital infrastructure coverage maps.

The background studies underpinning this book address a wide array of policy questions—on the barriers to DT adoption as well as on the effects of DTs on efficiency and equity—that could help support the jobs and economic transformation agenda in Senegal and beyond. This work begins to close significant knowledge gaps and suggests concrete policy recommendations to foster more inclusive technology upgrading. Key findings and insights could be explored by other low- and middle-income countries and allow for effective benchmarking. The work is a critical input into the forthcoming World Bank flagship report on the foundations of the digital economy for Africa, *Technological Transformation for Jobs in Africa: How Digital Can Support Inclusive Growth* (Begazo-Gomez, Blimpo, and Dutz, forthcoming) and could inform joint African Union and World Bank DE4A operational initiatives. The findings are a contribution to the ongoing policy discussion on the costs and benefits of expanding the coverage, access, and productive use of DTs by households and enterprises across the country and at the regional level. By showing the potential of DTs for enhanced firm performance and better jobs for more people, poverty reduction, and broader welfare gains, the work provides evidence and suggests recommendations that policy makers should consider when making decisions on the next wave of policy and program reforms. Finally, this book could inform operational engagements by the World Bank and other development partners aimed at fostering business development, entrepreneurship, and inclusion. It is hoped that the wide collaboration with government agencies in Senegal in the preparation of this book will provide a context for greater consensus-building and more effective multisectoral implementation during the COVID-19 recovery and beyond.

NOTES

1. As the PAP2/PSE states (in section II.2.1, "Challenges"), "the acceleration of structural transformation cannot be achieved without high productivity in high-growth sectors, massive job creation and an increase and diversification of exports. As such, it remains

fundamental to increase productive investments, to consolidate existing value chains and to implement, at the level of the regions, infrastructures to support development. It is equally important to accelerate the industrialization process, by relying on the development of SMEs, the promotion of national champions, the attraction of direct investments to capitalize on innovation opportunities, leveraging the potential of agricultural, tourism and mining, as well as the development of a new oil & gas ecosystem and a more innovative digital economy." The PAP 2A (Adjusted and Accelerated), spanning 2021–23, was modified in September 2020 and approved by the president on September 29, 2020. It is the latest plan that aims to implement the PSE initiated in 2012. The PAP 2A document states that the COVID crisis has brought new challenges related to the need for abundant, resilient, and high-quality agriculture; inclusive health; a high-performing education system; a strong local private sector; stronger social protection; and digital transformation.

2. See section I.3.3. of the PAP2/PSE.

3. In the PAP2, these are Costa Rica, Côte d'Ivoire, Kenya, Malaysia, Morocco, and Peru.

4. See para. 10, Senegal Numerique 2016–2025 (SN2025) strategy in appendix A.

5. Data are for 2017 and are from the World Bank's World Development Indicators database October 2020 edition. Internet penetration rate is defined as the number of individuals using the internet as a percentage of the population. Internet users are those individuals who have used the internet (from any location) in the past three months. The internet can be accessed via a computer, mobile phone, personal digital assistant, games machine, digital TV, and so on. Data are from the World Bank's World Development Indicators database October 2020 edition. Source organization: International Telecommunication Union, World Telecommunication/ICT Development Report and database.

6. Based on an exchange rate of US$1.00 = CFAF 555.45 in 2018. International Financial Statistics (database), International Monetary Fund, Washington, DC, https://data.imf.org/?sk=4c514d48-b6ba-49ed-8ab9-52b0c1a0179b.

7. Informal firms are defined in two ways: (a) In our work on larger firms (5 or more full-time employees), informal firms include all firms that do not use a standardized accounting system (according to ANSD, Senegal's National Statistical Agency's definition in its latest national enterprise census). (b) In our work on micro firms (where more than half of the sample are self-employed household enterprises with no full-time paid employees), informal firms are firms not having all the following indicators of formality: being registered with a local authority, being registered with the national revenue authority, paying local or municipal taxes, and being registered for VAT or sales tax.

8. The state of Ceará has about 9 million inhabitants and is ranked in the lower half of states in per capita income, 18th in a total of 27 states in Brazil, according to the Brazilian Institute of Geography and Statistics (IBGE) in 2019.

9. This book and most of the supporting background studies define firm size in terms of full-time paid employees: micro firms include self-employed firms with no employees up to 4–5 employees, small firms include 5–19 employees, medium firms include 20–99 employees, and large firms include 100 or more employees.

10. These tools include accounting and inventory control/point-of-sales (POS) software; the latter facilitates documenting and tracking the changing levels of inventories and customer purchases over time.

11. Including by the Agency for Development and Supervision of Small and Medium Enterprises (Agence de Développement et d'Encadrement des Petites et Moyennes Entreprises, ADEPME), Business Upgrading Agency (Bureau de Mise à Niveau des Entreprises, BMN), and General Delegation (Fund) for the Acceleration of Entrepreneurship (Délégation Générale à l'Entreprenariat Rapide, DER).

12. See chapter I.3.3., figure 6a, PAP2/PSE.

13. PREAC is the government's business environment and competitiveness reform program, which is embedded in the Plan Sénégal Émergent (PSE). The PREAC is entering its third phase, hence the abbreviation "PREAC3." References to the PREAC can be found in key PSE documents, such as PAP2/PSE.

14. See IFC (2020) *Creating Markets in Senegal: Country Private Sector Diagnostic*, which summarizes findings from prior analytical work, including the World Bank Group's *Senegal Enterprise Survey* (2014), the World Bank Group's *Doing Business* (2020), and the World Economic Forum's "Executive Opinion Survey" (2017). These priorities are also aligned with the PAP2/PSE highlighted in particular in section III.2.3. on reforms.

15. An interesting tool called CommAgri, the solution used to manage the Nataal Mbaye agricultural extension program, has shown promising uptake among producer collectives. This experience has inspired the development of a similar app, called Commango, which aims to better connect mango producer collectives to markets and (soon) to financing.
16. Data are from the 2018 Global Findex database.
17. This initiative is a pilot program initiated by Ecobank that seeks to equip thousands of merchants with virtual electronic payment terminals, which in turn allows Ecobank credit allocation based on payment records.
18. This survey was also administered in Ghana and Nigeria in western SSA; in Kenya, Rwanda, Tanzania, and Uganda in eastern SSA; and in Mozambique and South Africa in southern SSA.

REFERENCES

Bahia, Kalvin, Pau Castells, Genaro Cruz, Takaaki Masaki, Xavier Pedrós, Tobias Pfutze, Carlos Rodríguez-Castelán, and Hernan Winkler. 2020. "The Welfare Effects of Mobile Broadband Internet: Evidence from Nigeria." Policy Research Working Paper 9230, World Bank, Washington, DC.

Begazo-Gomez, Tania, Moussa P. Blimpo, and Mark A. Dutz. Forthcoming. *Technological Transformation for Jobs in Africa: How Digital Can Support Inclusive Growth*. Washington, DC: World Bank.

Cirera, Xavier, Marcio Cruz, Diego Comin, and Kyung Min Lee. 2021. "Firm-Level Adoption of Technologies in Senegal." Policy Research Working Paper 9657, World Bank, Washington, DC.

IFC (International Finance Corporation). 2020. *Creating Markets in Senegal: Country Private Sector Diagnostic*. Washington, DC: IFC.

Masaki, Takaaki, Rogelio Granguillhome Ochoa, and Carlos Rodríguez-Castelán. 2020. "Broadband Internet and Household Welfare in Senegal." Policy Research Working Paper 9386, World Bank, Washington, DC.

World Bank. 2014. *Senegal Enterprise Survey*. Washington, DC: World Bank. https://microdata .worldbank.org/index.php/catalog/2262.

World Bank. 2016. *World Development Report 2016: Digital Dividends*. Washington, DC: World Bank.

World Bank. 2020. *Doing Business 2020*. Washington, DC: World Bank.

World Economic Forum. 2017. "Executive Opinion Survey 2017: The Voice of the Business Community." In *Global Competitiveness Report 2017–18*, 333–39. Geneva: World Economic Forum.

1 Digital Technologies
ENABLERS TO "BUILD BACK BETTER"

THE IMPORTANCE OF DIGITAL TECHNOLOGY ADOPTION FOR TECHNOLOGICAL AND ECONOMIC TRANSFORMATION

Digitization can translate into increased economic opportunities for households and enterprises, including for the large number of low-skilled people living in low- and middle-income countries. Digital technologies (DTs) help reduce different types of costs, such as search, replication, transportation, tracking, and verification costs (Goldfarb and Tucker 2019). As costs decline, the shifts in economic behavior affect households' welfare, in part by affecting firms and governments. This book uses the term "economic transformation" to capture pathways to inclusive productivity growth, namely, better ways that generate more jobs and income, especially for low-income people. This book emphasizes technological transformation driven by innovation. Technological transformation as a pathway to more inclusive productivity growth occurs as enterprises adopt and more intensively use better existing technologies and/or leapfrog via the creation of new-to-the-world technologies—with these technologies generating more jobs and income for low-income people as well. In addition, complementary sources of inclusive productivity growth are spurred by policy and regulatory reforms and public investments. These reforms address the misallocation of resources—both *sectoral reallocation* of resources to more efficient, job-creating activities across firms and industries driven by market contestability, and *spatial integration*, that is, reallocation to more efficient, job-creating locations driven by different types of integration. The latter is driven by the following: greater regional regulatory harmonization, deeper regional digital platforms and trade, better links between smaller and larger local and global firms in specific value chains, urbanization and associated agglomeration economies, and improved integration across urban and rural/lagging regions.[1]

Evidence suggests that DTs can play an important role in facilitating technological and economic transformation for all enterprises, and for households, both as producers and as consumers. Reporting results from Argentina, Brazil, Chile, Colombia, and Mexico studies, Dutz, Almeida, and Packard (2018) show that low-skilled workers can also benefit from the more intensive use of the internet because of the output expansion effect from increases in productivity and consequent lower prices. Although firm-level use of a faster internet can result in a substitution effect, whereby some lower-skilled workers are replaced by the new technology, a sufficiently strong output expansion effect results in a net increase in the use of lower-skilled as well as higher-skilled labor.[2] More generally, as long as adoption of DTs expands production volumes and does not totally eliminate the need for lower-skilled workers, it results in more jobs over time, including both higher- and lower-skilled jobs. A large output expansion effect requires that demand be sufficiently responsive to the lower affordable prices, which is more likely in lower-income countries where demand for many products is still far from being satiated and therefore demand is more likely to be highly elastic (Bessen 2019). The output expansion effect is additionally facilitated through exports.

In Africa, the expansion of fixed broadband has been found to enable faster job creation and economic activity, though the greater potential lies with mobile broadband.[3] Most existing studies of DTs focus on basic cell phone access (or 2G technologies).[4] As for the internet, little is known about the causal effects of mobile broadband internet (for example, 3G/4G technologies) on the welfare of households and individuals. This lack of evidence is particularly troubling considering that most people in Africa access the internet through mobile phones rather than through fixed broadband internet.[5] A better understanding of the effects of mobile broadband on households as well as on different types of enterprises could have important implications for policy. A notable exception is a test of the effect of 3G/4G technologies on households, using data from Nigeria (Bahia et al. 2020). Those results show that the rollout of mobile broadband internet increased household consumption and contributed to reducing moderate and extreme poverty.

With its rising position as a West African tech hub, Senegal has significant potential to benefit from expanding access to the internet, which could contribute to increased labor and total factor productivity, agricultural production, and wage employment, as well as improved financial inclusion.[6] Over the past decade, Senegal has experienced a rapid expansion of digital technologies. Many Senegalese have gained access to basic cell phone services, and a sizable number also have gained access to mobile internet. Access to fixed broadband is growing but is still low. The number of incubators and investment funds in the country is also rising. Whereas Sub-Saharan Africa (SSA) is the region with the lowest internet penetration in the world (18.7 percent), Senegal has a relatively higher rate of internet penetration (29.6 percent), yet it remains well behind the global average (49.0 percent).[7] Figure 1.1 shows internet use and fixed broadband subscriptions for Senegal relative to peers, as well as to the 5th and 95th percentiles of SSA countries. Over the past 15 years, Senegal has ranked between first and second place in coverage compared with Côte d'Ivoire, Kenya, and Rwanda. Yet comparing Senegal to the regional leaders (figure 1.2) shows that the country still has room to catch up. Despite the rapid growth of digital infrastructure, Senegal still lags behind regional leaders in mobile broadband connectivity and internet use. A focus on unique mobile internet subscribers tells a similar story. Market penetration of mobile connections increased from 40 percent in 2011 to 52 percent in 2019, while the share of unique mobile internet subscribers increased from 11 percent to 31 percent. But again, as of 2019, the ratio of unique mobile internet subscribers was below that for leaders such as South Africa (50 percent) or Ghana (36 percent).[8]

Investing in expanding the digital infrastructure network in Senegal could foster growth and help reduce income inequality and poverty. A recent cross-country empirical study focusing on SSA shows the potential effects on economic growth and inclusion from an increase in the speed of the digital

FIGURE 1.1

Digital infrastructure in Senegal vs. peer countries

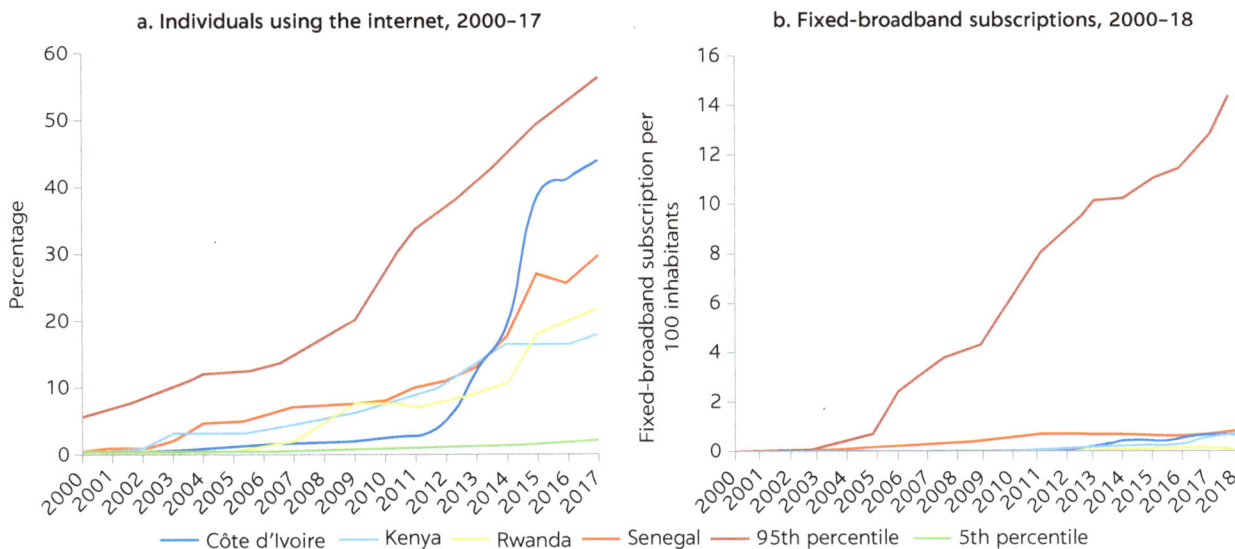

Source: World Bank staff, based on data from the World Bank's World Development Indicators database, with underlying data collected by the International Telecommunication Union, World Telecommunication/ICT Development Report and database. Accessed on October 2020.
Note: Red and green lines represent the values of countries ranking in the 95th and 5th percentile, respectively, of the cumulative distribution function (CDF) of the variable of interest for the group of countries that belong to SSA countries; the CDF uses an estimate of the latest year with available information. For more information about the methodology, see Mexico SCD (World Bank 2018). The 95th-percentile countries for individuals using the internet are Cabo Verde, Gabon, and South Africa, and the 5th-percentile countries are Burundi, Eritrea, and Somalia. For fixed-broadband subscriptions, the 95th-percentile countries are Cabo Verde and Mauritius, and the 5th-percentile countries are Chad, Democratic Republic of Congo, and South Sudan.

FIGURE 1.2

Mobile broadband and internet users in Senegal vs. leading regional countries

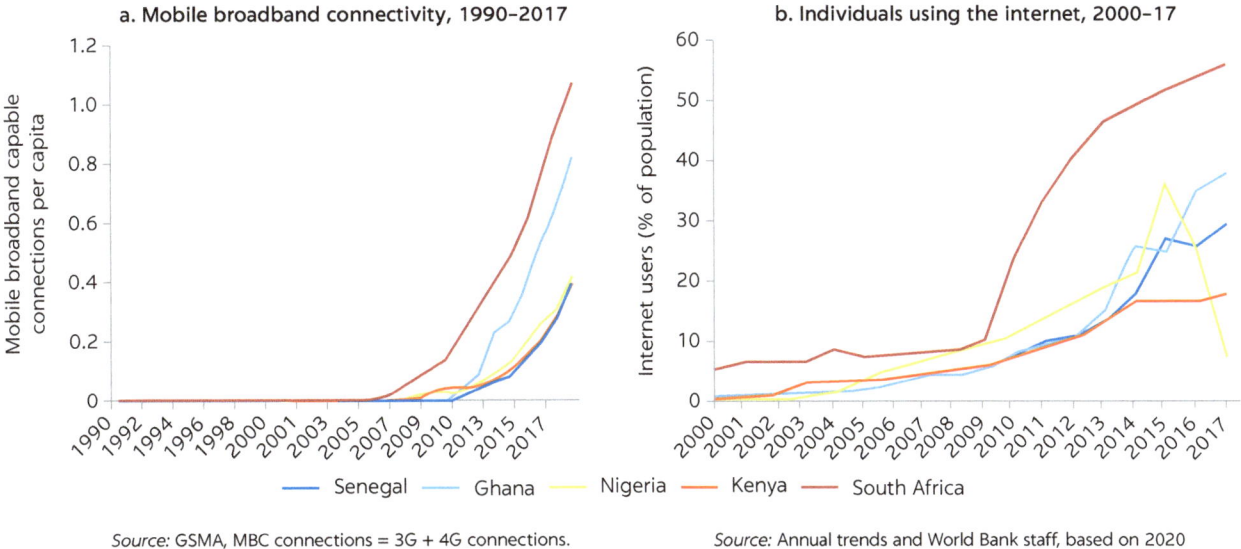

Source: GSMA, MBC connections = 3G + 4G connections.

Source: Annual trends and World Bank staff, based on 2020 World Development Indicators database.

infrastructure network expansion in Senegal relative to aspirational peers.[9] If the speed of network expansion in Senegal were to reach that of Chile or New Zealand (the 90th percentile of the world, excluding SSA), the rate of economic growth would increase by 1.7 percentage points per year, according to this study. Most of the effect would be through faster accumulation of capital per worker (2.6 percentage points). A similar counterfactual analysis was conducted for inequality and poverty. If the speed of network expansion in Senegal were again to reach that of Chile or New Zealand, inequality, measured by the Gini coefficient, would decline by 4 percent, and the poverty headcount index would drop by 2.2 percentage points. This analysis also concludes that the effects of digital infrastructure on economic growth appear to be larger for digital connections (3G and 4G) than analog ones (2G), and that there are significant complementary relationships between digital infrastructure and access to electricity from accelerating economic growth.

Over the past few years, the government of Senegal has taken several steps to create an enabling environment to digitalize its economy. Recently, the government implemented explicit reforms to increase internet usage, as outlined in "Stratégie Sénégal Numérique 2016–2025" (République du Sénégal 2016), and robust infrastructure is already in place. The new wave of sectoral reforms aims to further open the sector—including through the entry of more internet service providers[10]—and the consolidation and sharing of digital infrastructure, to catch up with regional leaders. Some indicators have already improved, such as in the speed of asymmetric digital subscriber line (ADSL) offers, and a reduction in the installation times of ADSL (from one month to five days).

Senegal is well connected to international networks through submarine fiber-optic networks. However, the distribution of international connectivity infrastructure across main telecom operators is highly uneven. Sonatel's access to the main international gateway "effectively gives it a monopoly over data transmission. The absence of regulatory measures focused on access to the international gateway further strengthens the dominant position of the historical operator and is likely to contribute to the high cost of international calls and data transmission in Senegal" (World Bank 2019, 34).

Senegal cannot make good use of its significant amount of fiber optics under public ownership. After sizable public investments, the State Informatics Agency (Agence de l'Informatique de l'État, ADIE) has a network of about 4,000 kilometers of fiber-optic cables. However, because of the technical, legal, and financial limitations of ADIE, the infrastructure is inefficiently managed. ADIE cannot provide access to its network because it does not have an operator's license. Beyond this legal barrier, ADIE is deemed to have insufficient capacity to manage its network efficiently (World Bank 2019).

Overall, broadband penetration and quality of service are low and the market is highly concentrated. Mobile markets in Senegal are highly concentrated and are dominated by Orange-Sonatel, which owns essential infrastructure. Until recently, monopolies in ADSL and 4G persisted.[11] The latter is partly a consequence of the opaque and ad hoc allocation to Orange-Sonatel of a 4G license on the basis of a bilateral negotiation rather than an open and competitive procedure.[12] The number of active mobile broadband capable subscriptions[13] in Senegal is above the average for the region (43.7 versus 35 per 100 inhabitants), as is the number of fixed broadband subscriptions (0.82 versus 0.43 per 100 inhabitants). Yet these numbers remain low compared with regional leaders; for example, the number of fixed broadband subscriptions in Namibia, Cabo Verde, and Mauritius (2.5, 2.8, and 21.6 per 100 inhabitants, respectively) is much higher than Senegal's. The country's low number of fixed broadband subscriptions is reflected in the fact that 98.5 percent of users in Senegal accessed internet through their mobile phones in 2018 compared with only 1.5 percent via fixed broadband.[14] Quality of service is low, despite access to three submarine cables and low wholesale prices. And broadband penetration lags behind regional leaders such as Cabo Verde, Côte d'Ivoire, Ghana, and Nigeria in indicators such as broadband density and internet bandwidth. Furthermore, the number of secure internet servers per 1 million people (17.1) is low compared with international standards (6,173 servers per 1 million people),[15] posing a security and governance challenge for internet adoption, e-commerce, and mobile money.

Digital divides remain in coverage and access to the internet across the territory and population groups. Despite significant improvement in coverage since 2016, mobile internet coverage and accessibility gaps across the country persist (see figure 1.3, panels a and b). The coverage rate of 2G in 2017 was 98.2 percent, while access to 3G was lower, at 78.2 percent. Despite relatively high levels of coverage, access to 3G services is low: 3G connections as a share of total market population were 26 percent in 2017, below the regional average of 30 percent.[16] In contrast, 2G connections as a share of total market population stood at 73 percent in the same year. In addition, an important 3G coverage gap persists between urban and rural areas, and there is significant connectivity inequality between Dakar and secondary cities (Pikine, Touba, and Saint-Louis), as well as within Dakar's neighborhoods. The country also has a relatively expensive data-only mobile broadband price basket, which can be a source of digital divide. In 2019, the cost to households—at 3.1 percent of its gross national income (GNI) per capita—was slightly above the levels established by the Broadband Commission for Sustainable Development at less than 2 percent of GNI per capita (ITU 2020).[17] The most vulnerable groups are more prone to be disconnected from any infrastructure except mobile phones (2G coverage is relatively high across both low- and high-poverty incidence areas).[18]

There are persistent horizontal inequalities in digital use, such as with mobile money. Senegal lags behind leading regional peers in mobile money account ownership, at 31.8 percent compared with 38.9 percent in Ghana and 72.9 percent in Kenya. There are also significant disparities within the country, for example, in relation to gender and location. Thirty-five percent of males own mobile money accounts in Senegal compared with 29 percent of women (figure 1.4, panel a),[19] while the share of mobile money account ownership in urban areas is 35 percent compared with 27 percent in rural areas (figure 1.4, panel b).[20] There are also gaps in usage by gender, age, education, income, and location in internet usage for digital payments, such as for purchases and bills. Online payments are more common among men, the more educated, those in the labor force, and people in the top 60 percent of the consumption distribution (figure 1.5). This limited use of digitally enabled financial services can be a barrier to profiting from digital sources of growth.

Although Senegal has traditionally lagged behind peer countries in DT adoption, it is catching up in digital financial service use. Figure 1.6 shows how Senegal's performance improved compared with benchmark countries between 2014 and 2017. By 2017, 39.5 percent of Senegalese households made or received digital payments, compared to 38.3 percent in Côte d'Ivoire and 38.9 percent in Rwanda. Yet Senegal is still significantly behind compared with best-performing countries such as Kenya, where 79 percent of households made or received digital payments in the past year. The same pattern holds once income is adjusted for the poorest 40 percent of the distribution. In terms of the share of households that used mobile phones to make payments, Senegal (at 33.5 percent) ranked just below Côte d'Ivoire (with 34.9 percent) and just above Rwanda (at 31.4 percent). When adjusting for the poorest 40 percent of the income distribution, Senegal (at 28.9 percent) fared better than Côte d'Ivoire and Rwanda. With

High market concentration and the persistent digital divide

a. Market concentration in mobile telephony and mobile internet, Senegal, 2016

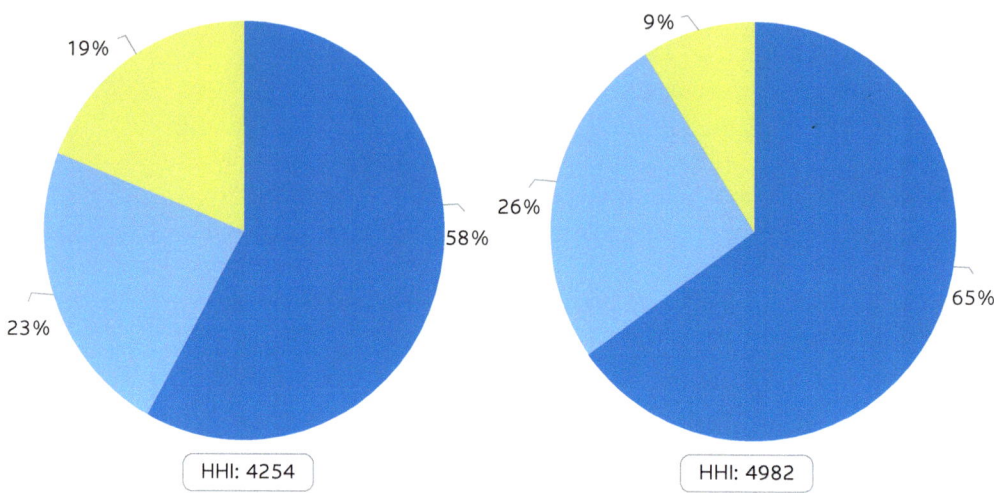

HHI: 4254	HHI: 4982
Market shares for mobile telephony by value	Market shares for mobile internet based on value

Source: World Bank DECA and competition report (2018), based on Autorité de Régulation des Télécommunications et des Postes (ARTP).
Note: HHI is the Herfindahl-Hirschman Index.

b. Mobile telephony coverage, Senegal, 2017

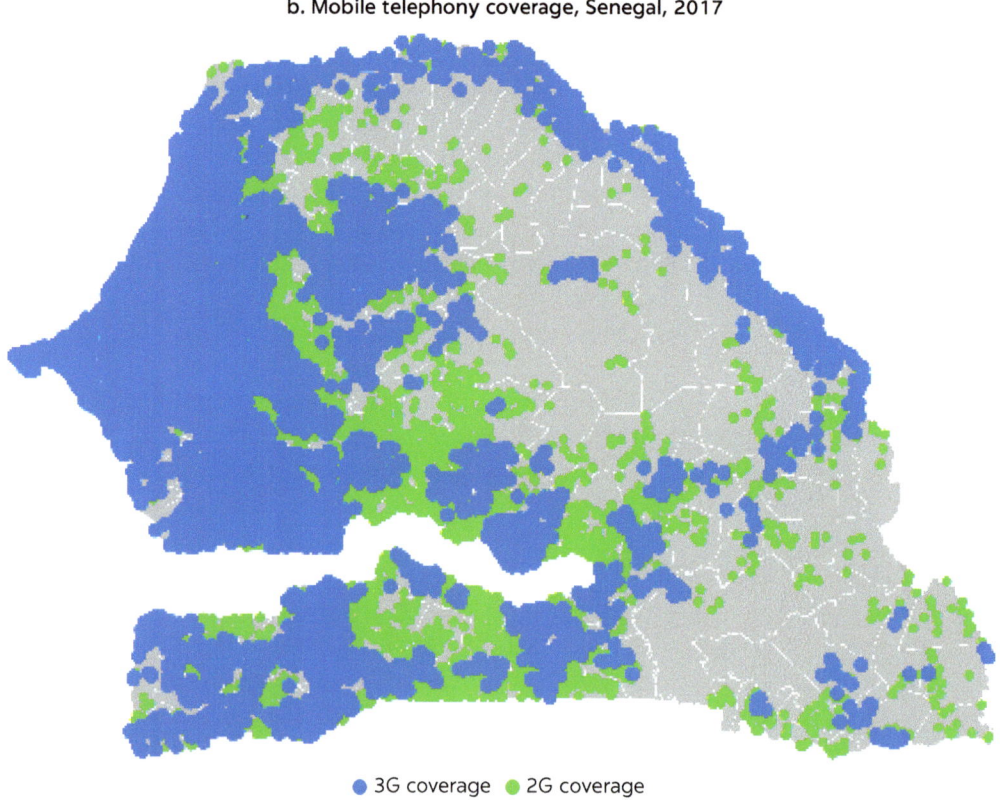

● 3G coverage ● 2G coverage

Source: World Bank staff elaboration based on Orange-Sonatel data.

FIGURE 1.4

Mobile money account ownership, West African Economic and Monetary Union countries, 2017

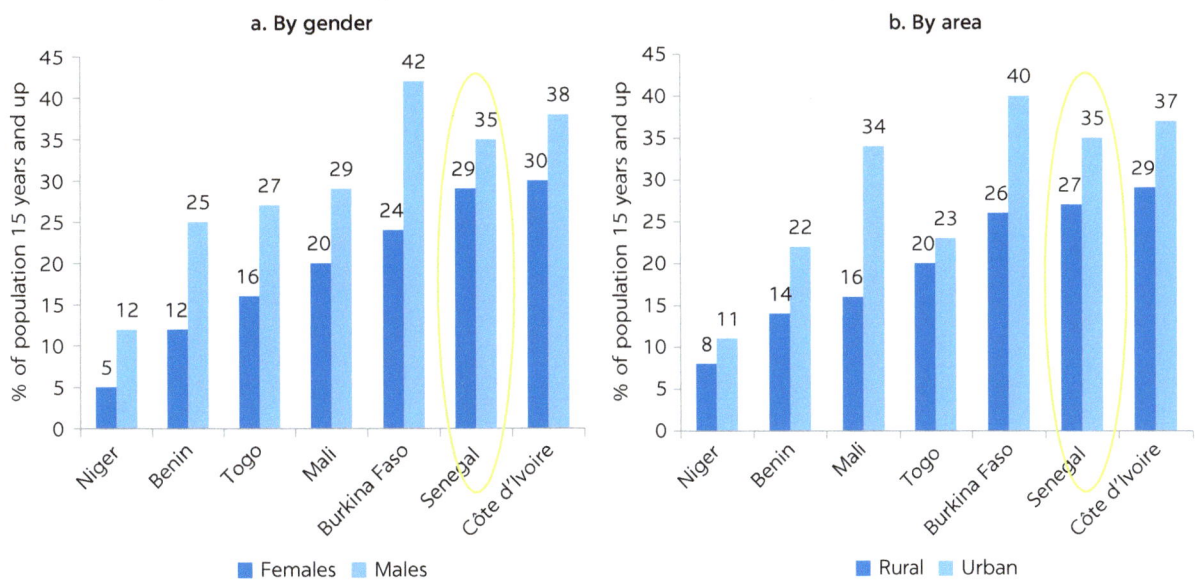

Source: Global Findex 2017 database.

FIGURE 1.5

Gaps in usage, by socioeconomic characteristics

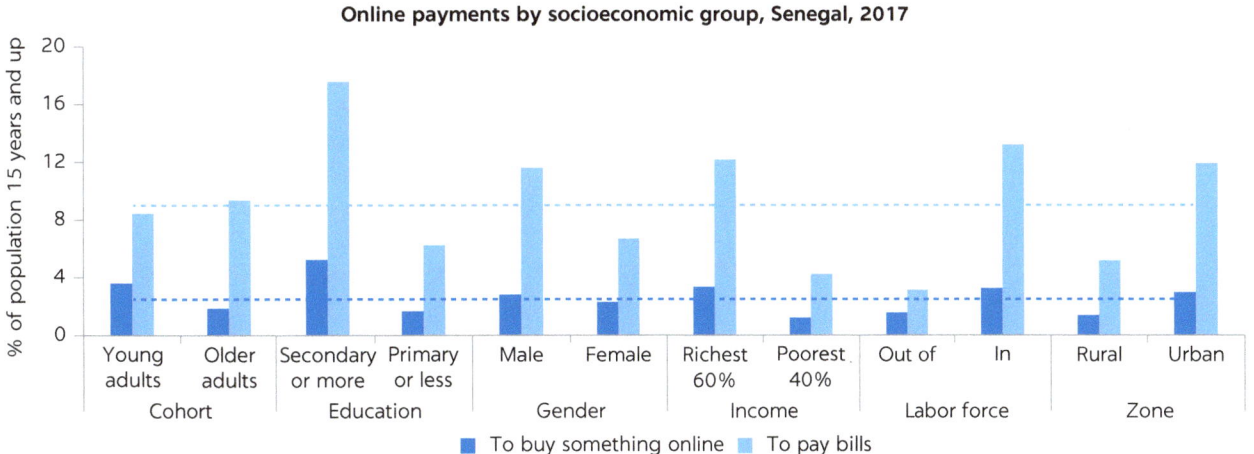

Source: Global Findex 2017 database.

regard to internet usage, Senegal also outperformed peer countries, with 10.4 percent of households using the internet to make payments or to buy something online, albeit significantly behind Kenya (at 26.1 percent). Adoption, however, does not directly translate to usage. In 2017, Senegal lagged behind Kenya, Rwanda, and Côte d'Ivoire in mobile money transactions per 100,000 adults. Though adoption by Senegalese households has improved in recent years, gaps remain in the magnitude of use.

This book aims to use new data and analyses to lead to a better understanding of the extent of digital and complementary technology adoption in Senegal—and how jobs and inclusion outcomes can be associated with this adoption to a

FIGURE 1.6

Digital services adoption by Senegalese households is catching up to peers

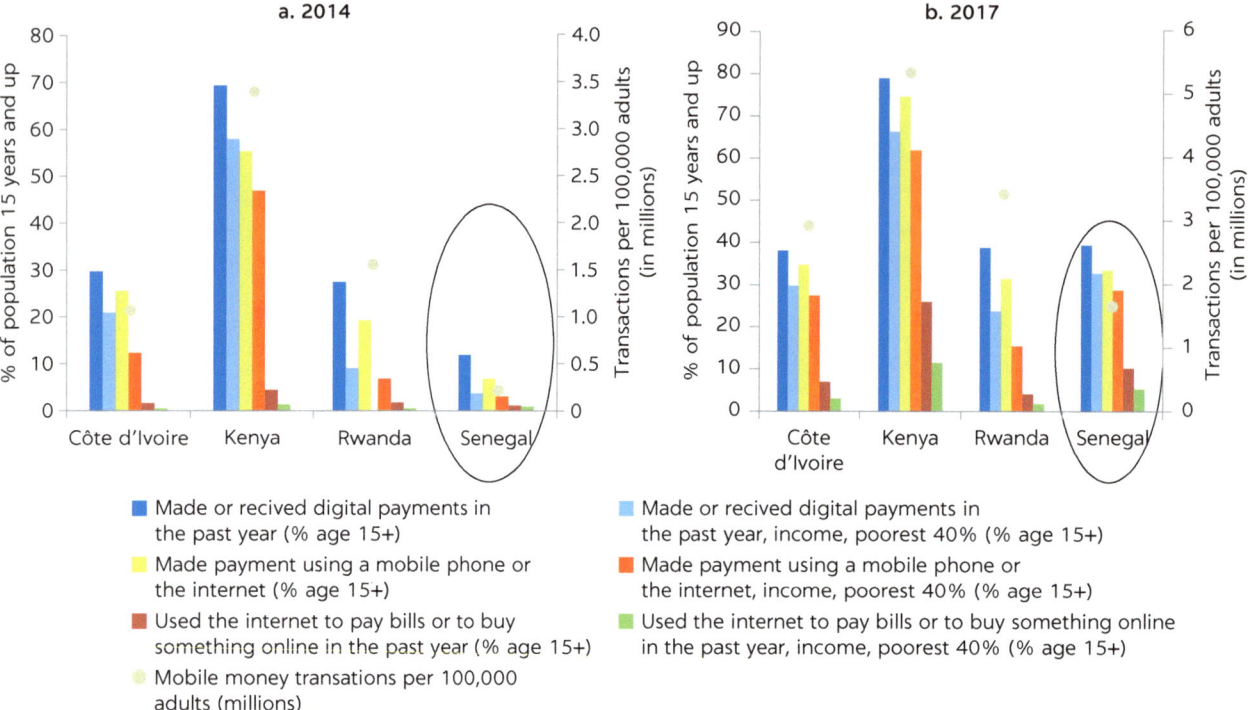

Made or recived digital payments in the past year (% age 15+)

Made or recived digital payments in the past year, income, poorest 40% (% age 15+)

Made payment using a mobile phone or the internet (% age 15+)

Made payment using a mobile phone or the internet, income, poorest 40% (% age 15+)

Used the internet to pay bills or to buy something online in the past year (% age 15+)

Used the internet to pay bills or to buy something online in the past year, income, poorest 40% (% age 15+)

Mobile money transactions per 100,000 adults (millions)

Sources: 2017 G20 Financial Inclusion Indicators from the World Bank's World Development Indicators database and the Global Findex 2017 database.

Note: "Made payment using a mobile phone or the internet" indicator for 2014 excludes the internet because of a change in the time series definition.

greater extent. To that end, the book summarizes its conceptual framework in figure 1.7. The framework links availability of DTs and their use to inclusive growth and household welfare. The framework builds on the theory of change proposed in the World Bank's Digital Economy for Africa (DE4A) initiative. It includes five enablers of DT availability—namely, infrastructure, skills, business, finance, and public platforms—facilitating use and impact. The framework begins with the figure's left column on availability of DTs. Affordable availability of both electricity and broadband internet infrastructure are necessary conditions as economywide enablers, without which none of the other more sophisticated DTs and complementary technologies that rely on electricity and the internet can be accessed and used. Affordable availability also includes investments in analog complements in addition to electricity infrastructure: the internet of things requires not only internet but also things, such as the tractors and irrigation systems on which data sensors can be installed. Policies are required to enhance access to affordable electricity, especially in poor rural areas, and to deepen ongoing reforms to boost competition in digital infrastructure and service provision to reduce consumer prices and improve the quality of services. Policies also are required to activate the Universal Services Fund (Fonds de Développement du Service Universel des Télécommunications, FDSUT) to reduce digital divides by leveraging private investments.

The middle column is about adoption and use, by firms, by individuals and households, and by government. Barriers to use by enterprises and individuals, in addition to the affordable availability of electricity and DTs on the supply side,

FIGURE 1.7

Conceptual framework: From DT availability to inclusive growth

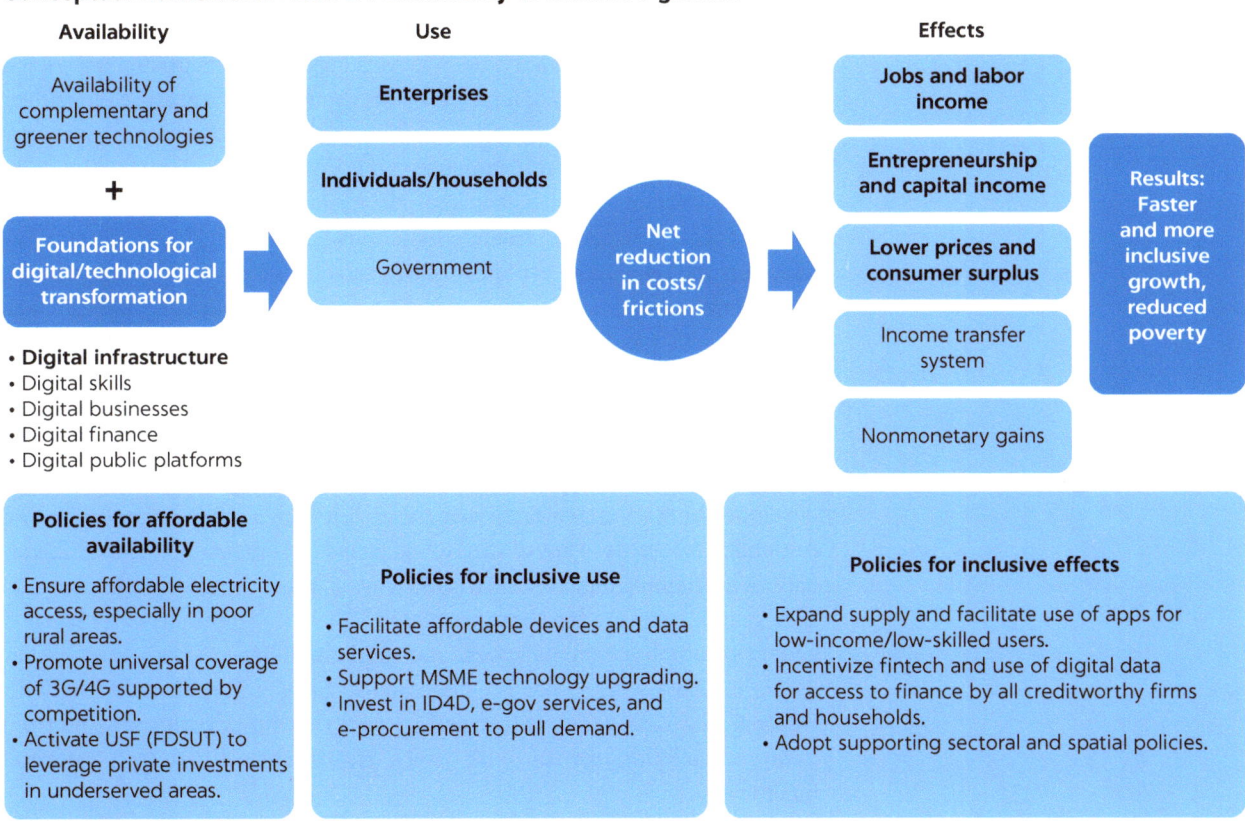

Source: World Bank.
Note: Text in bold represent the focus areas of this book.

stem from demand-side factors, including the following: purchasing power and access to finance; age, gender, language, and capabilities (owners/managers having vocational training for micro-sized firms); being part of effective networks such as having friends or other firms in one's ecosystem using DTs; and various types of risk and uncertainty. Addressing these demand-side factors may benefit from different types of usage incentives. Policies for more inclusive use include the following: facilitating affordable devices and data services, such as encouraging operators to offer zero pricing for basic messaging apps for poor people; promoting universal 3G/4G coverage; supporting technology upgrading by micro-, small-, and medium enterprises; and public investments to pull demand, including ID4D (digital identification systems to improve development outcomes), digital delivery of public services to households, and digitization of government-to-business transactions (such as access by all firms to tender for government procurement contracts).

Finally, as highlighted in the figure's right column, the effects of DTs depend on how intensively people use technologies that enhance productivity. Interactions with analog technologies and other complements as well as with the prevailing business environment also affect outcomes arising from DT adoption. The framework is centered on the distinguishing characteristic of DTs: their effect on reducing different types of economic costs or business-related frictions.[21] The internet is a "general purpose technology" that reduces costs across the economy and allows better data-driven decision-making,[22] which in turn can enable technological and broader economic transformation. The framework clarifies how these cost reductions have effects across five channels: jobs and labor income arising from lower costs faced by enterprises and individuals as workers; entrepreneurship and capital income earned by owners of larger firms and household enterprises; consumer surplus arising from lower prices, higher quality, and wider variety; the tax-transfer system; and nonmonetary gains.[23] As costs decline, the resulting shifts in economic behavior have implications for firms, households, and government. The cost-centric framework and its components highlight how DT adoption and use increase opportunities to access local and global product, labor, land, and financial markets by enterprises and individuals, as these costs include search, job-matching, transportation, and other transaction costs—clarifying as well that it is through the reduction of various costs that DT adoption and use facilitate business continuity when face-to-face or close-contact production of goods and services would otherwise be disrupted by COVID. Policies for inclusive effects include the following: incentivizing start-up software developers' creation and facilitating users' uptake of easy-to-use apps that respond to the needs of low-skilled workers and low-income, excluded individuals; strengthening digital MSME scoring systems to extend e-credits based on transaction records rather than on collateral (including for informal firms); and adopting complementary sectoral and spatial policies to ensure that investments in new technologies are allocated to firms and sectors in line with national comparative advantage and in ways that support the benefits from greater small-large enterprise links, rural-urban links, and regional integration.

The book's conceptual framework is further illuminated by a recent analysis of the mechanisms through which DTs affect poorer households' income-earning choices (Porto 2020). Porto uses household data from Senegal (and Kenya) to investigate the implications of lower consumer prices for rice and higher producer prices and lower input prices for groundnuts enabled by the adoption of specific DTs.[24] A DT upgrading that lowers consumer prices for rice benefits the average poor household twice as much as the average household, or the rural household more than the urban household, because rice is mostly consumed within the household and most rural households producing rice are net consumers. The welfare effects of a productivity improvement in groundnut production are significantly larger than those of a price increase at the farm gate (the price received by the producer from direct sales at the farm) because the productivity shock is assumed to be much larger than the price shock.[25]

The main chapters of this book—chapter 2, on households, and chapter 3, on enterprises—examine the extent of digital technology adoption, barriers to that adoption, and effects on outcome variables (depending on data availability), and they suggest policy options for greater inclusion supported by better jobs for more people. Chapter 2 focuses on digital infrastructure upgrading. It explores internet availability and uptake at the household level to identify its factors of adoption and effects on inclusive growth. The rest of the chapter follows this logic, looking at the main drivers (and barriers) of mobile internet adoption first, and then at its effect on welfare—including at the local level. The chapter finishes with a policy discussion that includes infrastructure-focused recommendations on broadening affordable internet access for all. Chapter 3 focuses on DTs and complementary technology upgrading by enterprises.

Its main message is that better jobs for more people require better and more firms. Better and more firms, in turn, require technology upgrading, more productive entrepreneurship, and more and better-allocated financing.

Vital aspects of digital technology adoption during the COVID-19 recovery

The relief, restructuring, and more productive recovery measures required as a response to the COVID-19 pandemic provide an opportunity to "build back better"—with digital technology adoption mattering even more than it did before. The COVID crisis has inverted the economic transformation agenda, which had been focused on generating better jobs for more people. It has refocused it over the coming months on relief measures to counter job destruction in the face of large declines in economic activity. These declines have been driven globally by disruptions in trade and value chains and reduced foreign financing flows (in the form of capital flight and lower foreign direct investment, foreign aid, remittances, and tourism revenues). Locally, social distancing measures by governments and citizens' responses have added to the reasons. In the new time of COVID-19, many of the productivity channels that were expected to be pathways to economic transformation are under threat, including financing for DT and other technology adoption, competition as a driver of sectoral reallocation, and trade and regional regulatory harmonization, better functioning supply chains, and local agglomeration economies as drivers of spatial integration. Also at threat are the reforms and investments needed for skills, infrastructure, and institutions to support inclusive productivity growth.

Moving forward, Senegal needs to adopt more radical incentive-driven reforms supported by an enhancement of capabilities to help it rebound forcefully following COVID-19. These reforms should include replacing existing rent-seeking structures and incentives with a focus on technology upgrading, entrepreneurship, and financing to build the skills and capabilities of unemployed and underemployed informal workers to engage in better work. Adoption of DTs and other complementary technologies matters even more, because it expands work and business opportunities, and together with entrepreneurship and financing support, boosts inclusive productivity growth.

Senegal needs better jobs for more people by focusing on both formal private firms and informal firms. Senegal has experienced an increase in joblessness. Its pre-COVID-19 growth acceleration (above 6 percent per year since 2014) failed to create enough jobs to meet its growing labor supply and reduce poverty sustainably. Senegal's key poverty-reduction challenge is to create more than 320,000 jobs each year, because higher earnings through jobs is the only sustainable way to reduce poverty. Senegal is rich in young people—Senegal is adding more than 300,000 people to its labor force each year. Over the current decade to 2030, Senegal will add more than 4 million people to its labor force.[26] The formal sector in Senegal in 2015 employed about 318,000 people, of which only about 188,500 were in the private sector (accounting for only 5 percent of the active working-age population), so Senegal needs to create more than all its current "formal sector jobs" each year to absorb new entrants into the labor force (figure 1.8). Therefore, creating better jobs for more people needs to include (a) better and more formal

FIGURE 1.8

Population in employment categories in Senegal, 2015

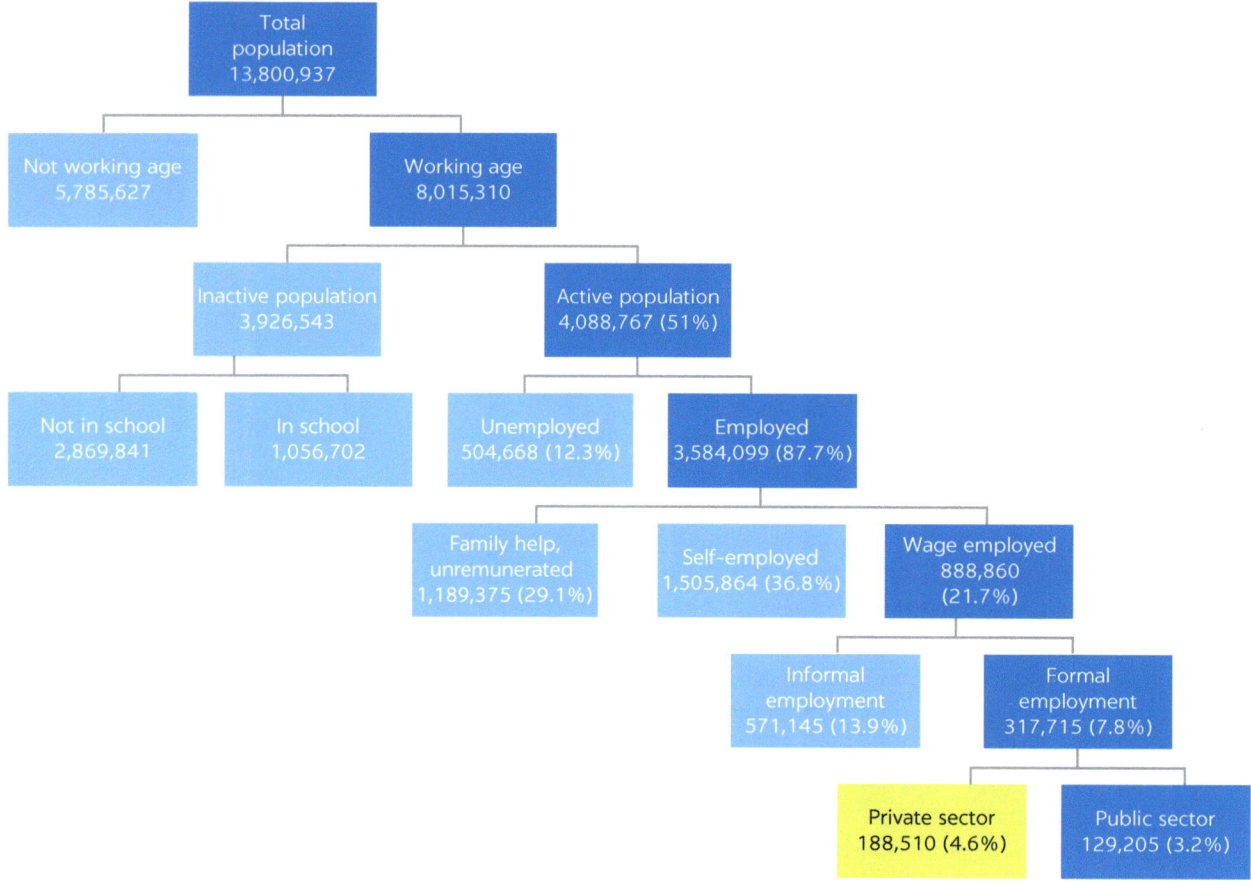

Source: Enquete Nationale sur l'Emploi au Senegal, Agence Nationale de la Statistique et de la Demographie, 2015.

private firms—to come closer to the technology frontier as well as leapfrog through the creation of new-to-the-world technologies, and to have more formal entrepreneurship; and (b) higher productivity for informal sector firms—for those able to take advantage of better government support services and other advantages of formalization, jump the quality hurdle, and join the modern sector in specific value chains.

Private businesses are facing new challenges since the COVID-19 outbreak.[27] Recent surveys RE conducted by the World Bank suggest that the downturn in sales has been large and widespread (figure 1.9a), with small firms being the most affected overall (figure 1.9b). Ninety percent of businesses experienced a decline in sales during the 30 days before the survey compared with the same period in 2019 during the late April–early May 2020 period, with slightly fewer firms though still a majority (76 percent) experiencing a decline in sales in the December 2020–early January 2021 period relative to a year earlier. Importantly, the drop was widespread across sectors, firm sizes, regions, and firm age categories. The estimated average reduction in sales for small firms was significantly larger compared with medium and large firms in the initial April–May 2020 period, with an average drop in sales of 55 percent compared with 40 percent for medium and large-sized firms, with a continued drop in sales of roughly 40 percent across all

FIGURE 1.9

Sales comparisons and predicted effect, by size of firm in the COVID-19 pandemic

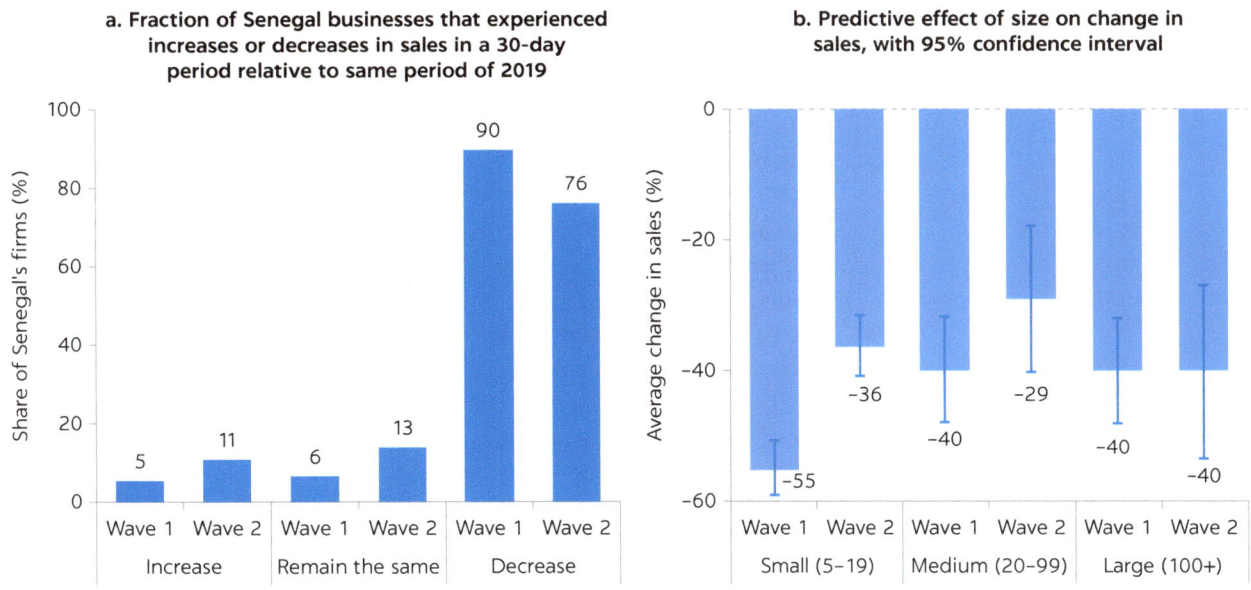

a. Fraction of Senegal businesses that experienced increases or decreases in sales in a 30-day period relative to same period of 2019

b. Predictive effect of size on change in sales, with 95% confidence interval

Source: COVID-19 Business Pulse Surveys in Senegal, April 28–May 8, 2020 (wave 1) and December 10, 2020–January 8, 2021 (wave 2).

FIGURE 1.10

Changes in digital technology uses by firms during the COVID-19 pandemic

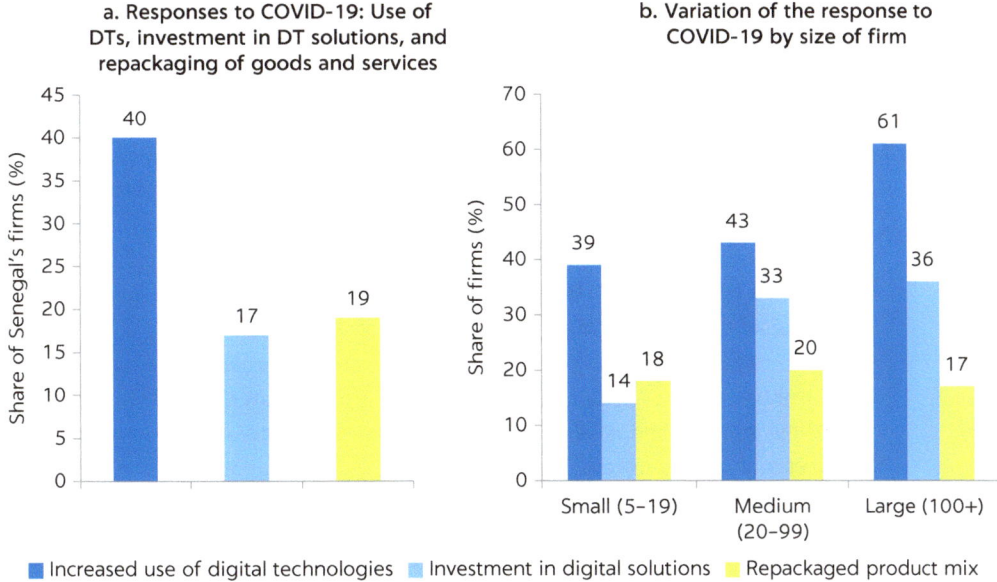

a. Responses to COVID-19: Use of DTs, investment in DT solutions, and repackaging of goods and services

b. Variation of the response to COVID-19 by size of firm

■ Increased use of digital technologies   ■ Investment in digital solutions   ■ Repackaged product mix

Source: COVID-19 Business Pulse Surveys in Senegal, April 28–May 8, 2020 (wave 1) and December 10, 2020–January 8, 2021 (wave 2).
Note: Because all three questions are "in response to COVID-19," only responses to the second wave of the survey are reported. However, for those firms in the panel (that responded to both waves), a "yes" response was recorded whether they replied affirmatively in either wave.

size groups in the December 2020–January 2021 period (though slightly less, at 29 percent, for medium-sized firms), controlling for other observable characteristics such as sector, age group, region, and exporting status.

The adjustment of firms to the crisis signals the importance of DT upgrading for businesses. In response to the COVID-19 outbreak, 40 percent of firms have started to use, or have increased the use of, the internet, online social media, specialized apps, or digital platforms for business purposes. About 17 percent of firms are investing in new equipment or software to adapt their product mix, while 19 percent of firms are changing their product mix. These results suggest that there already is a sizeable increase in demand for DT solutions applied for business, which can also lead to new opportunities for digital entrepreneurs in Senegal. An ongoing study is investigating if having access to DTs before the COVID-19 pandemic helped firms to cope better with this shock. Of greater concern in terms of widening digital divides is the finding that the share of medium and large firms investing in DT solutions in response to COVID-19, at 33 and 36 percent of these groups, respectively, is more than twice as large as it is for small firms, at 14 percent, as figure 1.10 shows. A similar divide exists between formal and informal firms. If these digital divides remain unaddressed, these diverging investment trends are expected to widen productivity, sales, and owner and worker income gaps over time.

DATA AND METHODOLOGICAL LIMITATIONS, KNOWLEDGE GAPS, AND FUTURE WORK

Several caveats and limitations were identified in the process of the preparation of this book. First, regarding the studies of digital adoption by households and individuals, there is scarce micro-level information on the usage of DTs such as fixed and mobile (2G/3G/4G) broadband internet, mobile money, and use of other digital platforms. For instance, the 2017–18 household budget survey, the Experimental Light Survey 2017–18 (ELEPS), [28] does not offer any direct or indirect measure of whether individuals/households use mobile broadband internet—which account for a large majority of internet users in Senegal—nor on ownership of smartphones. Even if household surveys ask questions on the use of internet such as in the EHCVM 2018/2019 survey, such self-reported internet access measures tend to be highly imprecise because of the lack of understanding by households and individuals of what access to the internet really means (as shown by International Telecommunication Union [ITU] and Research ICT Africa [RIA] studies). Because of these limitations, some proxy measures for internet use/adoption were used in the analysis, particularly to measure trends over time. However, these proxies are not free of measurement errors. Collecting more granular and reliable information on coverage, access, and use of the internet through micro-level surveys would allow a more rigorous analysis on barriers of adoption of new technologies and their effects on outcomes of interest.[29]

Expanding the availability of key data on coverage, access, and use of DTs is key for data integration and to conduct a more rigorous analysis of the importance of DTs for jobs and economic transformation. Second, the results presented in this book correspond to cross-sectional analyses only. Data with granular spatial coverage of digital infrastructure such as 2G/3G/4G were very difficult to obtain, sometimes only for some providers and with limited temporal coverage. Having reliable longitudinal data on digital infrastructure and the use of other DTs such as mobile money and digital platforms—which can be disaggregated at the subnational level—would allow future work to identify potential spatial gaps in access to DTs and how these gaps have changed over time. It also would allow a better understanding of the economic and welfare implications of such digital divides.

The analyses included in this book do not make any causal claims about the impacts on welfare and efficiency of DTs or provide exact evidence on what factors cause individuals or firms to adopt new technologies. Given limitations in the data available, the findings presented in this book should be interpreted as correlations and upper- or lower-bound effects of the relationship between DTs and outcomes. Identifying causal relationships requires rigorous implementation of empirical strategies based on quasi-experimental methods to explicitly address multiple issues such as self-selection in the dependent variable, survival bias, endogeneity, and measurement error, among other issues.

Several important policy questions linking DTs to household welfare and firm performance remain to be answered and should be the subject of future research. It would be desirable to continue expanding the evidence base on the efficiency and equity effects of DT adoption. A long list of questions remains unanswered and should be the subject of future work. For example, what is the relative contribution of different factors that explain the effect of DTs on welfare and poverty? What is the current and potential future role and effect of greater youth and women participation, at owner/entrepreneur, managerial, and worker levels, and across sectors?[30] How do education levels and digital skills affect the degree to which individuals benefit from access to internet and associated DTs? What are the critical complements that maximize the gains from DT adoption? How does the use of digital platforms such as mobile money and e-commerce affect efficiency and equity? Who would benefit most from a fast rollout of digitalization of government functions, particularly delivery of services such as the World Bank's ID4D (Identification for Development) initiative; asset registries including land, taxation, health and education; and social protection services that mitigate various risks by providing insurance? What are the business functions that are most relevant for inclusive productivity growth, and what is the relative importance of different DTs (such as availability of different digital finance services applications including insurance services and management applications such as inventory control and point-of-sale software)? Is the support of access and the capabilities to use internet and DT solutions that ride on the internet to upgrade general business functions sufficient to significantly improve productivity? Or is it also essential to support access and the capabilities to use the DT solutions to upgrade sector-specific business functions if sustained productivity growth and better jobs for more people are to be achieved?

Additional work to identify the causal effects of technology adoption on firm performance is being produced using the Firm-Level Adoption of Technology (FAT) survey (introduced in the overview). The first study explains the main sources of variation in technology adoption across countries, sectors, and regions, and within firms, and it shows how these variations are associated with productivity levels. This analysis compares Senegal with Brazil and Vietnam. The second study seeks to understand how technology adoption, especially DTs, can help firms to better cope with COVID-19. A follow-up survey already has been implemented in a subsample of firms in Senegal, Bangladesh, Brazil, and Vietnam. The third study categorizes the patterns of digitalization among firms to understand how investments in infrastructure and other potential complementary factors can further stimulate this process.

Further work also is being considered using the Business Pulse Survey to monitor the continuing effect of COVID-19 on businesses. Additional work is being explored with ANSD and ADEPME to conduct new waves of the firm-level survey and integrate Senegal's analysis with those done by other countries.

Senegal is the first country to have completed the first wave of this survey. Its analysis has been used as an example for other countries. Ongoing work includes benchmarking the data for cross-country comparisons.

Additional work on these issues would benefit from new data such as the Enquête Harmonisée sur les Conditions de Vie des Ménages (2018/2019) (EHCVM) survey, and a second round of the Senegal FAT survey. The EHCVM survey contains more granular information on the adoption and usage of DTs and asset ownership at the individual/household level. It allows better geographic integration of household survey data with digital infrastructure data. The EHCVM survey also allows for benchmarking digital adoption and use by households with regional peers, given that these data have been harmonized across WAEMU countries. A second round of the FAT survey in the form of a longitudinal instrument would allow implementing of quasi-experimental techniques to conduct more rigorous analyses of determinants of technology adoption at the firm level. It also would allow exploring the causal effects of technology upgrading on different dimensions of firm performance, including the contributing factors leading to better jobs for more people.

Finally, the implementation of a Public Expenditure Review (PER) of business support policies and programs is a priority. Such a PER, based on a methodology implemented in a variety of country contexts, is divided into three phases: (a) a mapping of policy instruments supporting entrepreneurship and innovation (both technology adoption as well as the generation and commercialization of new-to-the-world and new-to-Senegal technologies); (b) a functional analysis of the existing programs and policy instruments; and (c) an efficiency analysis. Close collaboration with the Ministry of Economy is ongoing to obtain more details on programs supporting businesses, with a focus on entrepreneurship as part of the entrepreneurship ecosystem assessment. Questionnaires have been implemented with some agencies, such as ADEPME and DER. However, this exercise has typically been more effective when implemented in collaboration with the unit responsible for budget prioritization and allocation (for example, Ministry of Finance). Members of the Ministry of Economy, including the private sector development unit, are very supportive of this exercise because it contributes to the rationalization of scarce fiscal resources and to the identification and elimination of the duplication of functions and activities across different government agencies. Ultimately, a PER of business support policies and programs should be an ongoing activity that helps policy makers understand the impact of expenditures and improve policies and programs over time—supported by a transparent process of "diagnostic monitoring," with new programs benefiting from structured experiments that are subject to monitoring in the form of diagnostics, namely, to learn what works and what does not, and to constantly raise the program's benefits by using new learning.[31]

NOTES

1. These three types of economic transformation and their different sources of inclusive productivity growth—technological, sectoral, and spatial transformations—require complementary policy reforms and public investments to expand productive private investment in ways that are inclusive, including skills, infrastructure (both soft infrastructure like finance and hard infrastructure, especially digital and energy as well as transportation and logistics) and institutions (especially to ensure macroeconomic stability and the required transparency and accountability for good governance). See Rodrik

and Sabel (2020) on the positive social externality of good jobs, and the significant economic, social, and political costs of failure to generate sufficient good jobs that private firms typically do not take into account.

2. The assumption that technology in combination with more skilled workers increases firm productivity is known as skill-biased technological change. This assumption is borne out in most of the country studies summarized by Dutz, Almeida, and Packard (2018). Even though the number of low-skilled jobs increases along with increased use of the internet, there is a relatively greater increase in the number of higher-skilled jobs. However, the Chile study, focusing on the adoption and use of complex software rather than the internet, finds an increase in the number of low-skilled production jobs over the six-year period analyzed, with the levels of employment of high-skilled production workers and managers not changing. Whether the effect of DT adoption on workers is inclusive is also found to depend on whether firms invest in workers. For an application of these ideas to Africa, see Choi, Dutz, and Usman (2020).

3. See Hjort and Poulsen (2019). The study also finds that higher-skilled workers benefit relatively more from fixed broadband expansion in the form of increased employment and earnings.

4. 2G technologies enable voice, SMS, and limited internet access, while 3G and 4G technologies enable faster internet browsing and data downloading, as well as other digital solutions such as picture and video taking and their uploading, and more specialized productivity-enhancing applications. Evidence shows that mobile phone expansion and coverage can have positive effects on household income, consumption, and poverty reduction (Beuermann, McKelvey, and Vakis 2012; Blauw and Franses 2016; Blumenstock et al. 2020). One of the mechanisms identified behind the positive effects is a rise in female employment associated with the rollout of mobile phone networks (Klonner and Nolen 2010). Mobile phones also improve access to information, reduce costs, and improve coordination, with positive effects on agricultural production, prices, and market access and participation (Aker and Mbiti 2010; Aker 2008, 2010; Aker and Fafchamps 2015; Jensen 2007; Muto and Yamano 2009; Zanello 2012). The literature also highlights the role of mobile money in improving access to the financial system: reducing transaction costs, easing the ability to make payments, and facilitating savings (Aker and Wilson 2013; Demombynes and Thegeya 2012; Munyegera and Matsumoto 2018).

5. The number of active mobile broadband subscriptions per 100 inhabitants in Africa in 2019 was 34, compared with 0.4 for fixed broadband subscriptions (ITU 2019).

6. Evidence suggests that the internet can contribute to firms' labor productivity and employment (Fernandes et al. 2019; Paunov and Rollo 2014). It has been associated with positive effects on output, crop prices, and wage employment in rural markets (Goyal 2010; Kaila and Tarp 2019; Ritter and Guerrero 2014; Salas-Garcia and Fan 2015). Mobile broadband has been linked with positive financial inclusion outcomes (Hasbi and Dubus 2019). Heterogeneous effects analyses suggest evidence of stronger labor income effects among new users and low-income households (De los Rios 2010; Marandino and Wunnava 2014) and improved labor market outcomes for women (Menon 2011; Chun and Tang 2018; Viollaz and Winkler 2020). E-commerce has been linked to gains in income among rural households (Couture et al. 2018) and with positive effects on reducing spatial inequality (Fan et al. 2018). Access to digital platforms such as mobile money—which are not exclusive to mobile broadband and are often based on 2G technology—can affect consumption and reduce poverty (Suri and Jack 2016), including by improving the distribution of government cash transfers (Aker et al. 2016).

7. Data are for 2017. Internet penetration rate is defined as the number of individuals using the internet as a percentage of the population. Internet users are those individuals who have used the internet (from any location) in the past three months. The internet can be accessed via a computer, mobile phone, personal digital assistant, games machine, digital TV, and so on. Data are from the World Bank's World Development Indicators database October 2020 edition and are based on the International Telecommunication Union, World Telecommunication/ICT Development Report, and database.

8. The source for these data is the GSMA. When consumers use multiple SIM cards to take advantage of special discounts or avoid high charges for off-net calls, market penetration in terms of unique subscribers provides a better picture of the degree of access to mobile services. The ratio of subscribers with multiple SIM cards in Senegal is reported to be around 55 percent and is expected to remain at that level over the next 4 years (based on email correspondence with Sonatel, June 15, 2020).

9. The study is based on a sample of 177 countries (47 of which are in SSA) from 1990 to 2018. The data are organized in nonoverlapping, 5-year panel data observations to avoid the influence of economic fluctuations. To address the issue of likely endogeneity and reverse causality, the study uses the GMM-IV system estimator. See Calderón and Cantú (2020).

10. In July 2018, three internet service providers were granted licenses, each with coverage obligations in five regions of the country. https://www.researchandmarkets.com/research /xr4gnn/the_senegal?w=4.

11. Tigo, Senegal's second-largest mobile network operator by number of subscribers, has been rebranded as Free Senegal as of October 1, 2019, following the company's acquisition by the Saga Africa Holdings consortium in April 2018. A website featuring the Free brand identity has been live since then and promotes the operator's new "4G+" LTE-A network, which is said to cover Dakar and most regional capitals and offer the country's fastest mobile data speeds.

12. See World Bank (2019, 33). According to TeleGeography, "a month after the Sonatel award, in July 2016 Tigo (now operating as Free) said it was preparing the groundwork for LTE and confirmed that it was in talks with the ARTP to secure a license, yet it was not until December 2018 that the ARTP granted a concession to Tigo for XOF27 billion," after the operator was taken over by a consortium (TeleGeography, GlobalComms Database, Senegal, December 2019, 9).

13. Estimations based on data from GSMA Intelligence. Accessed on June 2020. MBC connections = 3G + 4G connections. Regional calculation was done by adding 3G and 4G for all 48 countries in SSA and dividing by the total population. Eritrea was excluded because of a lack of available data for 3G and 4G connections.

14. According to the Internet Observatory of Senegal's Regulatory Authority for Telecommunications and Post (Autorité de Régulation des Télécommunications et des Postes, ARTP).

15. Data are from the World Bank's World Development Indicators database and refer to the number of distinct, publicly trusted TLS/SSL certificates found in the Netcraft Secure Server Survey. The sources for these data are Netcraft (http://www.netcraft.com) and World Bank population estimates.

16. Data are from the GSMA Intelligence 2020 database. Accessed on June 2020.

17. The data-only mobile broadband basket consists of a monthly data allowance of at least 1.5 GB, irrespective of the device used, over a 3G or higher network.

18. The lowest 2G coverage—at around 92 percent of the population—is in the Kédougou region, which has the second-highest poverty rate in the country, according to data from mobile services providers and the EHCVM 2018/2019 survey.

19. Senegal's gender gap—where ownership among males is about 21 percent higher than for females—is slightly larger than Côte d'Ivoire's but smaller than in Niger or Benin, where male ownership is more than twice the share of female ownership.

20. Rural clusters are defined as areas with a density of at least 300 inhabitants per km² of permanent land and with at least 500 and less than 5,000 inhabitants. Urban clusters are composed of towns and semidense areas and cities. Town and semidense areas are defined as clusters with a density of at least 300 inhabitants per km² of permanent land, a built-up surface share on permanent land greater than 0.03 percent, and at least 5,000 inhabitants. Cities are defined as clusters with a density of at least 1,500 inhabitants per km² of permanent land or with a built-up surface share on permanent land greater than 0.5 percent, and with at least 50,000 inhabitants.

21. Goldfarb and Tucker (2019) explore how standard economic models change as certain costs fall substantially and approach zero with the use of digital technologies. They focus on five types of lower costs: search (including matching suppliers and demanders), replication, transportation, tracking, and verification (with the final two jointly reducing monitoring costs).

22. General purpose technologies, or GPTs, are transformative technologies (such as the steam engine at the time of the Industrial Revolution in the late eighteenth century, the electric motor in the late nineteenth century, and the internet) that are characterized by pervasiveness (used as inputs by many downstream industries), inherent potential for technical improvements, and the creation of many positive spillovers. As GPTs are adopted across the economy, they generate economywide productivity gains. For a seminal article, see Bresnahan and Trajtenberg (1995).

23. Effects on the tax-transfer system include those linked to greater efficiency in tax collection and in the targeting and delivery of social transfers and longer-term public

investments, as well as from greater accountability from citizens for public domestic revenue mobilization and expenditures by enabling better governance and oversight of the collection and spending of public resources, and more transparent anticorruption measures. Effects on nonmonetary gains include those linked to quality-of-life-related benefits from better delivery of health, education, and other public services, as well as positive contributions to human dignity and individual empowerment.

24. The Kenya case study complements the Senegal explorations to illustrate the effect on nonfarm enterprises, many of which are informal businesses. They cannot be explored in the Senegal case because of lack of a detailed data.

25. Even though, as Porto (2020) argues, the price shock is more likely to reach a larger number of groundnut producers than a technology shock. For example, more exhaustive price information via cell phones would arguably benefit a larger number of farmers than advice on input use or on how to better utilize specific types of agriculture practices. Information can be used immediately if it is credible, but changing agricultural practices may require complementary skills or may need to circumvent cultural barriers. In the end, there will be a balance between the plausible size of the shock (7 percent price shock versus 32 percent technology shock) and its accessibility.

26. The share of youth younger than 20 makes up 53 percent of Senegal's population. Senegal needed to create 329,000 jobs to keep pace with net entrants into the job market in 2020, namely, the number of people age 15 (new entrants) minus those age 65 (exiters). This number is projected to rise to 427,000 for Senegal in 2030, and to 500,000 by 2050. Senegal's population of 16.7 million is expected to double to 33.2 million by 2050. These figures are based on the UN World Population Prospects 2019 standard projections (medium variant).

27. The COVID-19 Business Pulse Survey (COV-BPS) was implemented in Senegal during the last week of April and the first week of May 2020 in a first wave, and during end-December 2020 to early January 2021 in a second wave, to obtain information on the population of formal and informal firms with five or more employees. The implementation of the survey included establishments in Dakar, Thiès, Diourbel, Kaolack, Saint-Louis, and Ziguinchor. These regions account for 75 percent of businesses and 82 percent of employment in the 2016 establishment census. Surveyors conducted 508 phone interviews between April 28 and May 8, 2020 (wave 1), and 505 interviews between December 10, 2020 and January 8, 2021 (wave 2). In wave 2, 374 were panel firms (responding to both waves) and 131 were replacement firms. These samples are representative of the universe of formal and informal firms with five or more employees according to the Recensement Général des Entreprises (RGE), the latest establishment census conducted by the Agence Nationale de la Statistique et de la Démographie (ANSD). The samples were stratified across firm size based on number of employees (small 5–19, medium 20–99, and large 100+), sector, region, and formal status. The estimations include sampling weights to produce nationally representative indicators.

28. The 2017–18 Enquête Légère Expérimentale sur la Pauvreté (Light Experimental Poverty Assessment Survey, ELEPS), carried out by the Agence Nationale de Statistique et de la Démographie (ANSD), sampled three geographic areas across Senegal: Dakar, other urban centers, and rural areas, with an 89 percent response rate.

29. The RIA (Research ICT Africa) After Access Household and Individual survey 2017–18 also does not have a direct question on mobile internet subscriptions. However, "using mobile internet" can be defined based on two available variables, namely, the combination of "having a smartphone" and "using internet at least once a day." These data are very rich in the level of detail on usage questions asked. This data collection initiative, including both household/individual and complementary business surveys, have up until now been supported by Canada's International Development Research Centre (as well as the Swedish International Development Cooperation Agency). The initiative is invaluable for the World Bank Group's digital work and for policy makers in all African countries—especially its nationally representative, cross-country, and longitudinal features. Initiatives like these warrant sustainable data funding and Africa-wide institutional support, ideally in collaboration with national statistical agencies.

30. Table C.1 in appendix C highlights that Senegal has a relatively low number of firms with female owners: only 28 percent of firms covered by the FAT survey (a representative survey of firms with 5 or more workers) and 35 percent of firms covered by the RIA business survey (a representative survey of smaller micro firms, with most being either informal or semiformal)—relative to an average of 51 percent of RIA firms

across eight other SSA countries. Senegal stands out in having more female owners in micro agricultural firms (56 percent) but fewer in other sectors (manufacturing and services) relative to the other SSA countries. The female ownership variable did not have a statistically significant association with either adoption of smartphones or with firm productivity or profitability, in both FAT and RIA data for Senegal. More work is needed to better understand the role and effect of women and youth in firms' use of DTs, as well as in individual and household usage (including the use of distance learning for girls).

31. On the concept of diagnostic monitoring or "learning by monitoring," see Sabel (1994).

REFERENCES

Aker, Jenny C. 2008. "Does Digital Divide or Provide? The Impact of Mobile Phones on Grain Markets in Niger." BREAD Working Paper 177, Bureau for Research and Economic Analysis of Development (BREAD), http://ibread.org/bread/working/177.

Aker, Jenny C. 2010. "Information from Markets Near and Far: Mobile Phones and Agricultural Markets in Niger." *American Economic Journal: Applied Economics* 2 (3): 46–59.

Aker, Jenny C., and Kimberley Wilson. 2013. "Can Mobile Money Be Used to Promote Savings? Evidence from Northern Ghana." Swift Institute Working Paper 2012-003, SWIFT Institute, London.

Aker, Jenny C., and Marcel Fafchamps. 2015. "Mobile Phone Coverage and Producer Markets: Evidence from West Africa." *World Bank Economic Review* 29 (2): 262–92.

Aker, Jenny C., Rachid Boumnijel, Amanda McClelland, and Niall Tierney. 2016. "Payment Mechanisms and Antipoverty Programs: Evidence from a Mobile Money Cash Transfer Experiment in Niger." *Economic Development and Cultural Change* 65 (1): 1–37.

Aker, Jenny C., and Isaac M. Mbiti. 2010. "Mobile Phones and Economic Development in Africa." *Journal of Economic Perspectives* 24 (3): 207–32.

Bahia, Kalvin, Pau Castells, Genaro Cruz, Takaaki Masaki, Xavier Pedrós, Tobias Pfutze, Carlos Rodríguez-Castelán, and Hernan Winkler. 2020. "The Welfare Effects of Mobile Broadband Internet: Evidence from Nigeria." Policy Research Working Paper 9230, World Bank, Washington, DC.

Bessen, James E. 2019. "Artificial Intelligence and Jobs: The Role of Demand." In *The Economics of Artificial Intelligence: An Agenda*, edited by Ajay Agrawal, Joshua Gans, and Avi Goldfarb . Cambridge, MA: National Bureau of Economic Research; Chicago: University of Chicago Press.

Beuermann, Diether, Christopher McKelvey, and Renos Vakis. 2012. "Mobile Phones and Economic Development in Rural Peru." *Journal of Development Studies* 48 (11): 1617–28.

Blauw, Sanne, and Philip Hans Franses. 2015. "Off the Hook: Measuring the Impact of Mobile Telephone Use on Economic Development of Households in Uganda Using Copulas." *Journal of Development Studies* 52 (3): 315–30.

Blumenstock, Joshua, Niall Keleher, Arman Rezaee, and Erin Troland. 2020. "The Impact of Mobile Phones: Experimental Evidence from the Random Assignment of New Cell Towers." https://nkeleher.com/publication/blumenstock-keleher-rezaee-troland-forthcoming/.

Bresnahan, Timothy, and Manuel Trajtenberg. 1995. "General Purpose Technologies: 'Engines of Growth?'" *Journal of Econometrics* 65 (1): 83–108.

Calderón, César, and Catalina Cantú. 2020. "Assessing the Impact of Digital Infrastructure on Development in Sub-Saharan Africa." Internal report, World Bank, Washington, DC.

Choi, Jieun, Mark A. Dutz, and Zainab Usman, eds. 2020. *The Future of Work in Africa: Harnessing the Potential of Digital Technologies for All*. Africa Development Forum. Washington, DC: World Bank. https://openknowledge.worldbank.org/handle/10986/32124.

Chun, N., and H. Tang. 2018. "Do Information and Communication Technologies Empower Female Workers? Firm-Level Evidence from Viet Nam." Asian Development Bank Institute Working Paper 545, Tokyo.

Cirera, Xavier, Marcio Cruz, Diego Comin, and Kyung Min Lee. 2021. "Firm-Level Adoption of Technologies in Senegal." Policy Research Working Paper 9657, World Bank, Washington, DC.

Couture, Victor, Benjamin Faber, Yizhen Gu, and Lizhi Liu. 2018. "E-Commerce Integration and Economic Development: Evidence from China." National Bureau of Economic Research Working Paper 24384, Cambridge, MA.

De los Rios, Carlos. 2010. "Welfare Impact of Internet Use on Peruvian Households." Instituto de Estudios Peruanos, Lima.

Demombynes, Gabriel, and Aaron Thegeya. 2012. "Kenya's Mobile Revolution and the Promise of Mobile Savings." Policy Research Working Paper 5988, World Bank, Washington, DC.

Dutz, Mark A. 2018. *Jobs and Growth: Brazil's Productivity Agenda.* International Development in Focus. Washington, DC: World Bank. https://openknowledge.worldbank.org/handle/10986/29808.

Dutz, Mark A., Rita Almeida, and Truman Packard. 2018. *The Jobs of Tomorrow: Technology, Productivity, and Prosperity in Latin America and the Caribbean.* Washington, DC: World Bank. https://openknowledge.worldbank.org/handle/10986/29617.

Enquête Harmonisée sur les Conditions de Vie des Ménages (EHCVM) 2018/2019. Agence Nationale de la Statistique et de la Demographie.

Fan, Jingting, Lixin Tang, Weiming Zhu, and Ben Zou. 2018. "The Alibaba Effect: Spatial Consumption Inequality and the Welfare Gains from E-commerce." *Journal of International Economics* 114: 203–20.

Fernandes, Ana, M. Aaditya Mattoo, Huy Nguyen, and Marc Schiffbauer. 2019. "The Internet and Chinese Exports in the Pre-Ali Baba Era." *Journal of Development Economics* 138: 57–76.

Goldfarb, Avi, and Catherine Tucker. 2019. "Digital Economics." *Journal of Economic Literature* 57 (1): 3–43.

Goyal, Aparajita. 2010. "Information, Direct Access to Farmers, and Rural Market Performance in Central India." *American Economic Journal: Applied Economics* 2 (3): 22–45.

Hasbi, Maude, and Antoine Dubus. 2019. "Determinants of Mobile Broadband Use in Developing Economies: Evidence from Sub-Saharan Africa." Working Paper hal-02264651, HAL.

Hjort, Jonas, and Jonas Poulsen. 2019. "The Arrival of Fast Internet and Employment in Africa." *American Economic Review* 109 (3): 1032–79.

IFC (International Finance Corporation). 2020. *Creating Markets in Senegal: Country Private Sector Diagnostic.* Washington, DC: IFC.

ITU (International Telecommunications Union). 2019. *Measuring Digital Developments: Facts and Figures 2019.* Geneva: ITU.

ITU (International Telecommunications Union). 2020. *Measuring Digital Development ICT Price Trends 2019.* Geneva: ITU.

Jensen, Robert. 2007. "The Digital Provide: Information (Technology), Market Performance, and Welfare in the South Indian Fisheries Sector." *Quarterly Journal of Economics* 122 (3): 879–924.

Kaila, Heidi, and Finn Tarp. 2019. "Can the Internet Improve Agricultural Production? Evidence from Viet Nam." *Agricultural Economics* 50 (6): 675–91.

Klonner, Stefan, and Patrick Nolen. 2010. "Does ICT Benefit the Poor? Evidence from South Africa." Proceedings of the German Development Economics Conference, Hannover.

Marandino, Joaquin, and Phanindra Wunnava. 2014. "The Effect of Access to Information and Communication Technology on Household Labor Income: Evidence from One Laptop per Child in Uruguay." IZA Discussion Paper 8415, Institute for the Study of Labor, Bonn.

Masaki, Takaaki, Rogelio Granguillhome Ochoa, and Carlos Rodríguez-Castelán. 2020. "Broadband Internet and Household Welfare in Senegal." Policy Research Working Paper 9386, World Bank, Washington, DC.

Menon, Nidhiya. 2011. "Got Technology? The Impact of Computers and Cell Phones on Productivity in a Difficult Business Climate: Evidence from Firms with Female Owners in Kenya." IZA Discussion Paper 5419, Institute of Labor Economics (IZA), Bonn.

Munyegera, Ggombe K., and Tomoya Matsumoto. 2018. "ICT for Financial Access: Mobile Money and the Financial Behavior of Rural Households in Uganda." *Review of Development Economics* 22: 45–66.

Muto, Megumi, and Takashi Yamano. 2009. "The Impact of Mobile Phone Coverage Expansion on Market Participation: Panel Data Evidence from Uganda." *World Development* 37 (12): 1887–96.

Paunov, Caroline, and Valentina Rollo. 2015. "Overcoming Obstacles: The Internet's Contribution to Firm Development." *World Bank Economic Review* 29 (suppl. 1): S192–S204.

Porto, Guido 2020. "Digital Technologies and Poorer Households' Income Earning Choices in Sub-Saharan Africa: Analytical Framework and a Case Study for Senegal." Internal report, World Bank, Washington, DC.

République du Sénégal, Ministère des Postes et des Télécommunications. 2016. "Stratégie Sénégal Numérique 2016–2025." Dakar.

Ritter, P., and M. Guerrero. 2014. "The Effect of the Internet and Cell Phones on Employment and Agricultural Production in Rural Villages in Peru." Working paper, University of Piura, Piura, Peru.

Rodríguez-Castelán, Carlos, Samantha Lach, Takaaki Masaki, and Rogelio Granguillhome Ochoa. 2021. "How Do Digital Technologies Affect Household Welfare in Developing Countries? Evidence from Senegal." Policy Research Working Paper 9576, World Bank, Washington, DC.

Rodrik, Dani, and Charles F. Sabel. 2020. "Building a Good Jobs Economy." In *Political Economy and Justice*, edited by Danielle Allen, Yochai Benkler, and Rebecca Henderson. Chicago: University of Chicago Press. http://www2.law.columbia.edu/sabel/papers/Building%20 a%20Good%20Jobs%20Economy%20November%202019_final.pdf.

Sabel, Charles F. 1994. "Learning by Monitoring: The Institutions of Economic Development." In *Handbook of Economic Sociology*, edited by Neil Smelser and Richard Swedberg, 137–65. Princeton, NJ: Princeton University Press and Russell Sage Foundation. Also in *Rethinking the Development Experience: Essays Provoked by the Work of Albert O. Hirschman,* edited by Lloyd Rodwin and Donald A. Schon. Washington, DC: Brookings Institution and Lincoln Institute, 231–74.

Salas Garcia, Vania B., and Qin Fan. 2015. "Information Access and Smallholder Farmers' Selling Decisions in Peru." Paper presented at the Agricultural and Applied Economics Association (AAEA) & WAEA Joint Annual Meeting, paper no. 205380, July 26–28, San Francisco.

Suri, Tavneet, and William Jack. 2016. "The Long-Run Poverty and Gender Impacts of Mobile Money." *Science* 354 (6317): 1288–92.

Viollaz, Mariana, and Hernan Winkler. 2020. "Does the Internet Reduce Gender Gaps? The Case of Jordan." Policy Research Working Paper 9183, World Bank, Washington, DC.

World Bank. 2014. *Senegal Enterprise Survey.* Washington, DC: World Bank. https://microdata .worldbank.org/index.php/catalog/2262.

World Bank. 2016. *World Development Report 2016: Digital Dividends.* Washington, DC: World Bank.

World Bank. 2018. *Mexico: Systematic Country Diagnostic.* Washington, DC: World Bank Group.

World Bank. 2019. "Poverty and Equity GP." Technical paper, Washington, DC, World Bank.

World Bank. 2020. *Doing Business in 2020.* Washington, DC: World Bank.

World Economic Forum. 2017. "Executive Opinion Survey 2017: The Voice of the Business Community." In *Global Competitiveness Report 2017–18,* appendix C, 333–39. Geneva: World Economic Forum.

Zanello, Giacomo. 2012. "Mobile Phones and Radios: Effects on Transactions Costs and Market Participation for Households in Northern Ghana." *Journal of Agricultural Economics* 63 (3): 694–714.

2 Households
WELFARE EFFECTS OF DIGITAL TECHNOLOGIES

This chapter aims to shed light on the main drivers of mobile internet adoption and on its effect on household welfare in Senegal, and then to offer policy recommendations to broaden affordable internet access for all.[1] The first step in understanding the potential effects of digital technologies (DTs) on welfare is to find out how households decide to adopt them or not, identifying the factors and barriers that drive behavior. The next step is to untangle their effect on welfare—first, for those who have access to the internet in general; and second, whether benefits are geographically differentiated and whether lagging regions are catching up. The focus of this chapter is on digital infrastructure, particularly internet access, and on households, to identify the factors for the adoption of DTs and their effects on welfare. The chapter also explores some of the channels of transmission, specifically, the effect of internet provider competition on welfare through a (lower) price of services and the resulting entry of new customers into the market; the role of the labor market as a potential channel of transmission toward improvements in welfare; and the role of mobile money in expanding transfers.

DRIVERS OF INTERNET ADOPTION

As access to the internet—in the form of fixed and mobile broadband—continues to expand in Senegal, it is increasingly important to understand the differences, if any, on which households and individuals decide to adopt these technologies.[2] A better understanding of these issues can inform the discussion of policies to bridge the digital divide. Despite important progress to expand internet access, there is evidence of a digital divide in SSA compared with the rest of the world (Calderón et al. 2019). Within countries, the gap is even broader, particularly in rural areas. And rural households in Africa, which are poorer on average, have been found to face lower rates of internet access (World Bank 2019).

The analysis on the determinants of internet adoption for households is based on a nationally representative household expenditure survey.[3] Specifically, the study exploits 2019 household survey data from the West African Economic and Monetary Union (WAEMU), namely, the Enquête Harmonisée sur les Conditions de Vie des Ménages (2018/2019) (EHCVM). The results presented in this section focus on the case for Senegal, where a total of 7,157 households were successfully interviewed, representing 36,754 individuals above the age of 15. The study analyzes the decision to adopt the internet considering the different mechanism of accessing the technology, with a primary focus on mobile internet adoption. This focus is particularly important in African countries, where most people access the internet through mobile phones rather than through fixed broadband internet.[4] Mobile internet adoption considers those individuals that accessed the internet through their mobile telephone devices in 2018, which for Senegal stood at 34.2 percent. Demographic and socioeconomic characteristics—such as age, gender, educational attainment, labor force status, language, as well as asset ownership variables (including age of cell phone, access to electricity, and whether the household owns a television set, a computer, and a tablet)—are considered. The analysis also considers policy-related variables, such as the price of mobile data and access to electricity.[5] Different internet access modalities—such as access to internet at home, work, school, university, cybercafes, and public places—are also incorporated into the analysis as covariates to determine if they complement or substitute for mobile internet adoption.

Significant connectivity gaps remain in income, gender, location, and education. Before looking at the probability of adoption, the analysis presents descriptive statistics by individuals with access to mobile internet in the country in 2019 (table 2.1).[6,7] Individuals living in households above the median income threshold are more likely to be connected to mobile internet (52 percent) than

TABLE 2.1 **Mobile internet connectivity data for individuals**

INDIVIDUALS CONNECTED AND NOT CONNECTED TO THE INTERNET THROUGH THEIR CELL PHONES

	SENEGAL	
DISTRIBUTION GROUPS	**CONNECTED (%)**	**NOT CONNECTED (%)**
Income threshold		
Above	52	48
Below	21	79
Urban	49	51
Rural	17	83
Men	40	60
Women	30	70
Age		
15 to 24 years old	37	63
25 to 40 years old	45	54
41 and older	21	79
Education		
Less than primary	20	80
Primary and secondary	48	52
Tertiary	86	14
Read and write French	54	46

(continued)

TABLE 2.1, *continued*

INDIVIDUALS CONNECTED AND NOT CONNECTED TO THE INTERNET THROUGH THEIR CELL PHONES

	SENEGAL	
DISTRIBUTION GROUPS	**CONNECTED (%)**	**NOT CONNECTED (%)**
Do not read and write French and illiterate	19	81
Sector		
Agriculture	13	87
Industry	45	54
Services	46	54
Unemployed/inactive	31	68
Household assets		
Computer ownership	65	35
Television ownership	47	53
Tablet ownership	62	38
Complementary infrastructure		
Access to electricity	47	53
Total population	34	66

Source: EHCVM 2018/2019.
Note: This table includes only individuals 15 and older and only those individuals who accessed the internet through their mobile phones. Each tabulation is carried out at the individual level, adjusting for household weights. The agriculture sector includes jobs in crop yields, fisheries, and animal breeding. The industry sector includes extractive, manufacturing and other industries, and public works/construction jobs. The services sector includes commerce, restaurants/hotels, transportation, communication, education, health, other, and personal services jobs. Secondary education is defined as individuals with less than tertiary but more than primary education. Tertiary is defined as individuals with tertiary education or more.

those below (21 percent). Connectivity is also highly correlated with education levels: 86 percent of individuals with tertiary education and above are likely to be connected, compared with only 20 percent with less than primary education. Being literate and proficient in the main language available online also plays a role. More than half of individuals who are able to read and write in French are connected to mobile internet compared to 19 percent of those who are unable to read and write in French or are illiterate. The sector of employment matters: 46 and 45 percent of workers, respectively, in the services and industry sectors are connected, compared with only 13 percent of those employed in agriculture. About a third of people who are unemployed or not participating in the labor force are connected, which could reflect individuals using the internet to look for jobs. Individuals between 15 and 25 years old were 37 percent likely to be connected, and this share increases to 45 percent for those between 25 to 40 years of age, while only 21 percent of people 41 years and older had access to mobile internet. With regard to gender, only 30 percent of women were connected to mobile internet, that is, 10 percentage points lower than men. Individuals who reside in rural areas are much less likely to be connected to mobile internet (17 percent) than those in urban areas (49 percent). Finally, household assets also play a role. More than 60 percent of individuals who live in households that own computers and tablets were connected to mobile internet.

Results from the probabilistic model of adoption[8] show that the most important factors for mobile broadband adoption are household income (measured by consumption per capita), price, gender, age, tertiary educational attainment, language, and employment sector, television ownership, access to electricity, and living in an urban setting.[9] Age follows a concave function, where individuals younger than 25 years are 15 percentage points more likely to have access to mobile internet than those older than 40 (the group of reference), while the likelihood of connection is even higher for individuals ages 25 to 40, who are 21 percentage points more likely to have access compared with the group of reference. With regard to household income, on average increasing monthly per capita expenditures by CFAF 40,000 (about US$72) per year would increase adoption by 9.3 percentage points.[10] Regarding education, high-skilled individuals (having tertiary education or more) are 16 percentage points more likely to adopt mobile internet. On language, individuals who can read and write in French are 13 percentage points more likely to access the internet through their mobile phones, suggesting that the language of content matters. Having access to the internet at work is also an important factor, increasing the probability of adoption by individuals by 19 percentage points. On other locations of access, having access to the internet at home has a complementary effect on the adoption of mobile internet (17 percentage points); having internet access through cybercafes is a direct substitute, reducing the probability of adoption by 13 percentage points.

Important barriers to adoption remain, including in relation to price, gender, location, and access to electricity (see figure 2.1). On average, a decline in the price of monthly mobile internet by CFAF 1,100 (about US$2.00) would increase adoption by 2.0 percentage points.[11] Being a woman lowers the likelihood of adoption by 6 percentage points compared with men. This gender gap, despite high levels of existing infrastructure, draws attention to the existence of barriers, which may limit adoption by women and other vulnerable groups. Urban location increases mobile internet adoption by 4.0 percentage points, reflecting the existence of a rural/urban divide. Furthermore, households with access to electricity are 6 percentage points more likely to adopt mobile internet. This result highlights the limitation that lack of electricity represents for households and the need to include its access in policies to encourage internet adoption. Another interesting finding highlights a link with the labor market, where employment in the service and industry sectors is associated with an increase of 8.1 and 9.4 percentage points in adoption, respectively.

The second part of the analysis on the determinants of internet adoption for households reiterates the importance of income, wealth, and having electricity, while also underlining the importance of education and network effects. This complementary analysis is based on another household dataset by Research ICT Africa (RIA), a representative dataset on internet use consisting of 1,233 observations for Senegal in 2017–18.[12] This analysis examines the correlates of mobile internet adoption (defined as a dummy variable that equals 1 if the respondent has a smartphone and uses the internet a least once a day). Both income and wealth (measured by the ownership of consumer durables such as automobiles, refrigerators, and television sets), having electricity, and having a job are important correlates of adoption (figure 2.2). Another important correlate is education: an increase in schooling from 5 to 15 years is associated with an increase in the probability of adoption from 4.5 percent to 12.3 percent, at the mean of the sample. The analysis also finds sizable network effects linked to having friends who use smartphones: an increase from 1 to 5 in the number of friends who use

FIGURE 2.1

Determinants of mobile internet adoption in Senegal, 2018–19

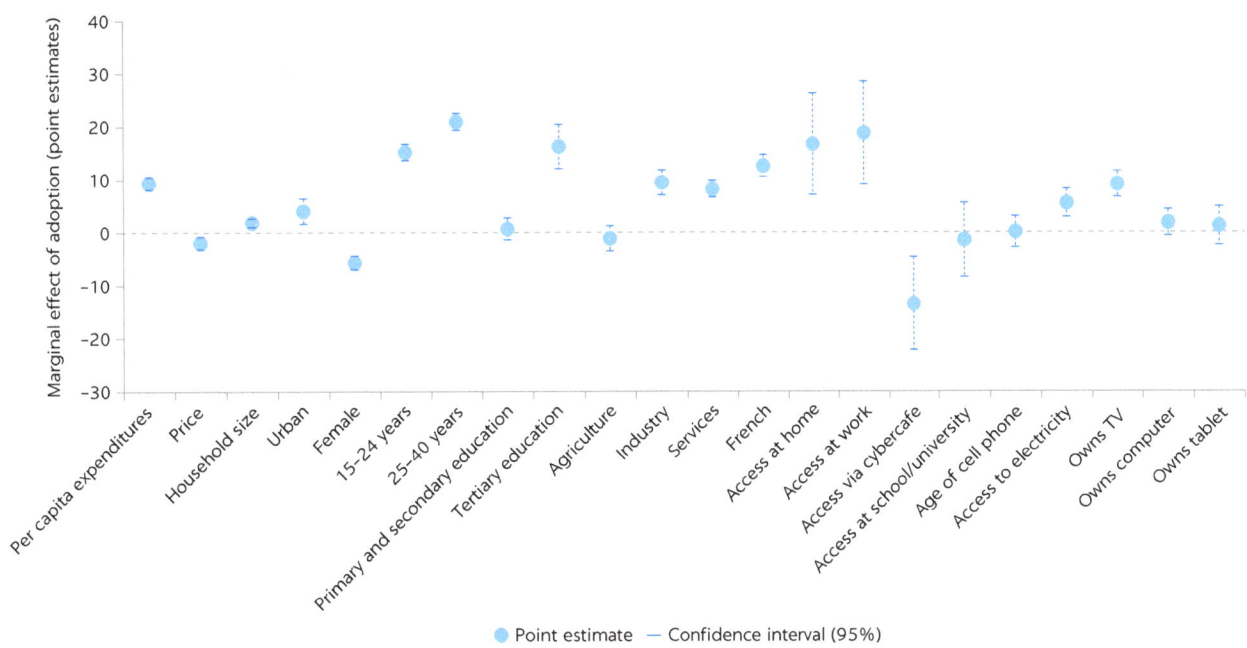

● Point estimate — Confidence interval (95%)

Source: Rodríguez-Castelán et al. 2021.
Note: Point estimates are at a 5 percent confidence interval from a Heckman-corrected probit model. Average marginal effects: Marginal effects for log per capita expenditure, price, and household size are calculated based on a one-unit increase in standard deviation, equivalent, respectively, to CFAF 40,000 (US$72.00) per capita per month, CFAF 1,100 (US$2.00) per month, and about six household members. Mobile internet access refers to when individuals access the internet through their mobile telephone devices. Mobile internet price is obtained by calculating the median expenditure of prepaid mobile phone cards and airtime/data transfers among mobile internet users in each country's geographic area in which the survey is representative. This value is then computed as a share of total consumption at the same geographic level to adjust for cost of living. This value is then imputed to each individual observed in the microdata. The baseline dummy for location refers to rural areas. The age baseline dummy is 41+ years. The base variable across education categories is individuals with less than primary education. Secondary education is defined as individuals with less than tertiary but more than primary education. Tertiary is defined as individuals with tertiary education or more. The base category for read/write French refers to national languages, other languages, and those that cannot read and write. Age of cell phone is the median value of time that the household has owned the device at the enumeration area level. The base variable across labor market sectors refers to inactive and unemployed. The agriculture sector includes jobs in crops, fisheries, and animal breeding. The industry sector includes extractive and other industries, and public works/construction jobs. The services sector includes commerce, restaurants/hotels, transportation, communication, education, health, other, and personal services jobs. Per capita expenditures, household size, access to electricity, and owning a computer, tablet, and television are household-level variables. All results are statistically significant, with the exception of primary and secondary education, individuals employed in the agricultural sector, median age of cellphone, those with access to the internet at school or a university, and households that own a tablet or computer.

messaging applications is associated with an increase in the probability of adoption from 2.5 percent to 37 percent.

IMPROVEMENTS IN WELFARE FOR HOUSEHOLDS WITH INTERNET ACCESS, INCLUDING THROUGH ITS EFFECT ON THE LABOR MARKET

Identifying the effects of the internet on welfare (and its transmission channels) is increasingly important as access continues to expand in Senegal. In a recent study, Hjort and Poulsen (2019) find positive effects of fixed broadband on employment rates across the skill distribution, with high-skilled workers benefiting the most, according to data from 12 African countries. Very little evidence exists, however, on the effects of mobile broadband on households and individuals, despite many people in low- to middle-income countries relying on it as their main way to access the internet. An important recent addition to the research,

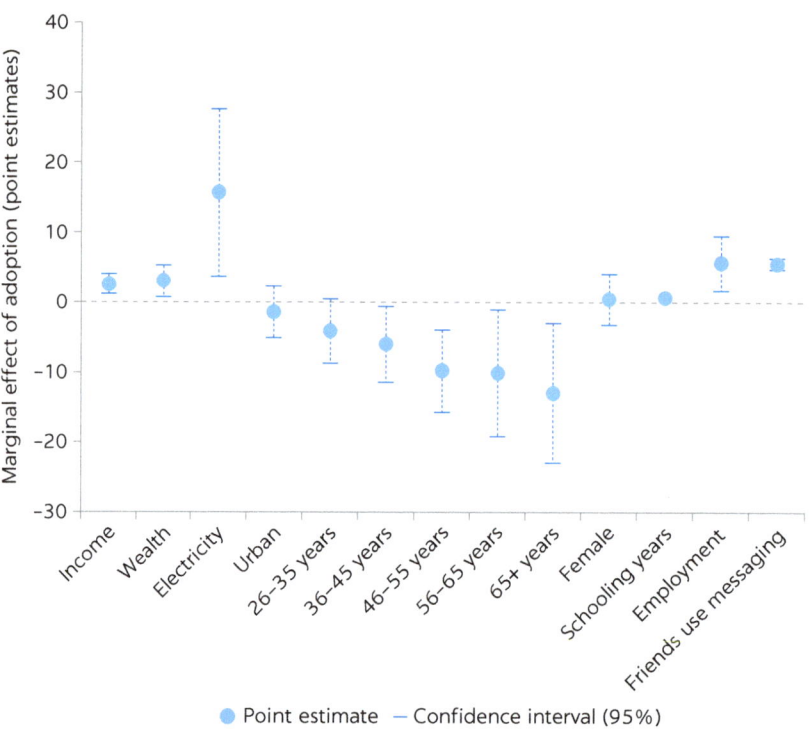

FIGURE 2.2

Main factors associated with mobile broadband internet adoption in Senegal

● Point estimate — Confidence interval (95%)

Source: Atiyas and Doğanoğlu 2020.
Note: Based on RIA 2017–18. Income is the total flow of earnings of the individual in a month, including wages and salaries, business income, property income, and so on. Wealth is a categorical index, with a range of 0–3, measuring if a person owns a refrigerator, a TV, and a car.

however, is the research of Bahia et al. (2020), who show that the rollout of mobile broadband internet increased household consumption between 7 and 11 percent and reduced extreme poverty between 5 to 8 percentage points in Nigeria.

To analyze how internet access correlates with poverty and household consumption, this section relies on a new study merging household survey data and mobile (and fixed) broadband internet coverage information.[13] The study considers the Experimental Light Survey ESPS 2017–18 (ELEPS) and 2017 mobile internet coverage and fixed-broadband infrastructure maps. The study matches the location of households—based on GPS coordinates of enumeration areas—with data on the location of fiber-optic transmission nodes and 2G/3G mobile coverage maps.[14] Coverage maps were obtained from Collins Bartholomew (as in the previous analysis), as well as from the main telephone communication providers: Senegal Expresso, Orange (its local subsidiary, Sonatel), and Tigo. The study performs a diagnostic of how proximity to—and coverage of—fixed and mobile broadband infrastructure affected household consumption and poverty in 2011 and 2017–18. It is one of the first studies to measure the effects of both mobile broadband and fixed-broadband infrastructure on households' welfare. The focus on mobile broadband is key because, as mentioned, most people access the internet through mobile phones in Senegal (ARTP 2019). The study is also the first to use rich data from Senegal's household consumption surveys to measure the welfare implications of different types of technology: mobile phone access and fixed and mobile broadband internet.

Mobile broadband internet coverage in Senegal is associated with higher household consumption and lower poverty rates.[15] These results showcase the potential benefit that access to mobile broadband internet can have on increasing consumption and reducing poverty in Senegal (figure 2.3). Specifically, estimates show that total consumption in households covered by 3G is 14 percent higher than that of households without coverage, conditional on a set of indicators that proxy for household wealth, education, and location. This figure is even larger for nonfood consumption, which is about 26 percent higher among 3G-covered households. Also, the findings show a 10 percent lower extreme poverty rate for households covered by 3G. To put this in context, similar work for Nigeria—integrating household panel data and historical coverage maps—finds an 11.1 percent increase in total consumption and a 7.9 percentage point decline in extreme poverty after three years of coverage (Bahia et al. 2020).[16] The analysis of heterogeneous effects in Senegal also shows differences in the magnitude of welfare effects across groups. Although the welfare effect of 3G coverage is evident in both urban and rural areas, its magnitude is larger in urban areas and for young or male-headed households.

Internet access has been positively associated with employment. Literature on the subject shows that the internet can lead to improvements in labor productivity and employment (Fernandes et al. 2019; Paunov and Rollo 2015). The literature has also found positive effects of the internet on agricultural markets (Goyal 2010; Kaila and Tarp 2019; Ritter and Guerrero 2014; Salas Garcia and Fan 2015). There is evidence of heterogeneous effects, with stronger labor income effects among new users and low–income households (De los Rios 2010; Marandino and Wunnava 2014) as well as of improved labor market outcomes for women (Chun and Tang 2018; Menon 2011; Viollaz and Winkler 2020).

FIGURE 2.3
Impact of 3G coverage on consumption and poverty in Senegal

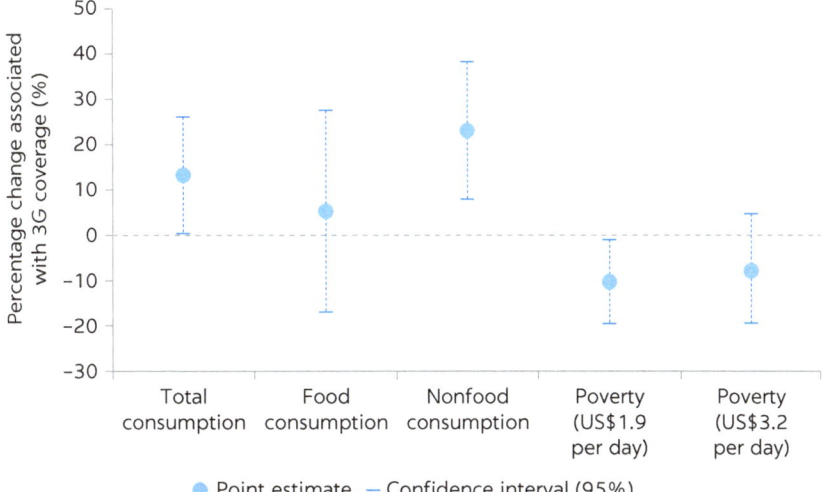

Source: Masaki et al. 2020.
Note: Point estimates at 5 percent confidence intervals. Ordinary least squares results across consumption type and poverty. Standard errors are clustered by enumeration areas. Additional controls include (a) at household-level, size, marital status, literacy, and sex of household heads, household access to electricity and housing conditions (as measured in a composite index of dwelling characteristics); (b) spatial controls such as elevation, nighttime light luminosity, and road density; and (c) urban/rural and region dummy variables.

Mobile broadband has also been linked with positive financial inclusion outcomes (Hasbi and Dubus 2019).

The study finds evidence of a labor market mechanism through which mobile broadband internet could be translating into improvements in welfare. The expansion of digital infrastructure and access to the internet may not only help the creation of jobs in the information and communication technology (ICT) sector but also reduce transaction costs for people in finding jobs, for productive inputs, or for improved labor productivity (World Bank 2016). The effect of mobile broadband technologies on wage/salaried or formal employment is of particular interest because a shift away from informal self-employment toward more productive wage/salaried or formal employment in private and public services is deemed to be a potential pathway to reducing poverty rates in Africa (World Bank 2016, 2019b). The study finds that 3G coverage is positively correlated with employment in "better" jobs—or formal, wage/salaried jobs with higher earnings.[17] These findings are consistent with other studies (for example, Bahia et al. 2020; Hjort and Poulsen 2019) showing similar results, namely, that access to the internet translates into increased employment in wage or higher-skilled jobs. The finding draws attention to the potential role that DTs can play in improving labor outcomes— and, particularly, employment in "better" jobs.

PRODUCTIVE USE OF THE INTERNET: THE CASE OF EXPANDING ACCESS TO MOBILE MONEY

The internet can facilitate access to markets and information to households, both as producers and consumers, contributing to an expansion of their welfare. From a consumer perspective, the internet is often employed as a source of news, to engage in e-commerce and do research on purchases, for paying utilities online, and for accessing government services such as telemedicine or health information. From an income-generating activity view, the internet can provide market-related information to farmers, enable small business transactions, and aid in job searches. Importantly, the internet can facilitate access to the financial system, benefiting both consumers and producers. It also enables people to communicate with each other, through email and chat, as well as to engage in social media. The internet can also be a platform for leisure activities, such as online games.[18]

Expanding mobile money in Senegal could help reduce horizontal inequalities in digital financial services and provide access to capital to a larger share of the population. Doubling the current share of mobile money users in Senegal from 31.8 percent to 63.6 percent—in alignment with one of the main targets of a World Bank operation for WAEMU—can have significant welfare implications, according to a simulation exercise (see box 2.1). Expanding access to mobile money accounts can reduce horizontal inequality in the use of mobile money, such as by gender, location, or income. Specifically, women's ownership increases from 29 percent to 57 percent, access to accounts in rural areas increases from 27 percent to 62 percent, and the mobile money gap between the poorest and richest income quintiles is reduced by 5 percentage points, according to simulation results. Improved access promoted by lower transaction costs can lead to an estimated 70 percent increase in remittances transactions, and twice the private payments related to labor activities. In the long run, increased access to digital financial services can have a positive effect on consumption and poverty reduction.

ACCESS TO DTs CAN PROMOTE LOCAL ECONOMIC ACTIVITY

As internet access continues to expand, it is valuable to find out whether its benefits are geographically differentiated and whether the more lagging regions are playing catchup.[19] This section presents the results of a recent analysis on the subject at the territorial level in Senegal.

The analysis investigates the potential benefits and costs of DTs at the local level using spatially disaggregated data.[20] The study takes advantage of spatially disaggregated information on poverty, a proxy for economic growth—that is, nighttime lights (luminosity)—and data on 2G coverage and fiber-optic network maps. Specifically, it integrates two poverty maps for 2005 and 2013 (produced following the method by Elbers, Lanjouw, and Lanjouw 2003) with information

BOX 2.1

Expanding digital financial services in WAEMU to improve livelihoods and reduce horizontal inequalities

DTs have the potential to improve livelihoods through different channels of transmission, an important one of which is increasing financial inclusion among previously underserved individuals. Policies that expand mobile financial inclusion can have significant benefits, including for groups of the population whose members have socioeconomic disadvantages, according to a recent simulation exercise for countries in the WAEMU region. Per this analysis, expanding financial inclusion can lead to a reduction in differences in account ownership by gender, education, rural/urban area, income, and labor force participation status. In this way, expanding the mobile money supply can have important effects on reducing inequalities between socioeconomic groups.

Expanding digital financial services can increase the provision and affordability of financial services to traditionally unserved and underserved populations in three ways: (a) fostering interconnectivity between banks; (b) reducing transaction costs; and (c) increasing the supply of mobile money agents to rural areas. This set of policies is part of a series of reforms that WAEMU is promoting to expand digital financial inclusion across the region.

The simulation exercise referenced here estimates the likely effects of expanding the supply of financial services with a focus on mobile money services in the WAEMU region. Although financial inclusion in WAEMU countries has increased over time, particularly for mobile money subscribers, the share of account holders is still relatively low and concentrates at the top of the welfare distribution. Clear subgroups of the population face a disadvantage in access to mobile money accounts: women, individuals with relatively low educational attainment, those who are out of the labor force, those living in rural areas, and those who are relatively poorer.

The analysis reveals that policies aiming to expand access to mobile money accounts to achieve the targets of the World Bank WAEMU's Financial Inclusion Regional Development Policy Operation Project would reduce subgroup differences, including greater take-up by women and those in the bottom 40 percent of income distribution. In addition to narrowing the gap on mobile accounts, expanding the access and use of mobile money can enhance welfare, including through increasing savings and investment in human capital, diversification of economic activities, and

continued

Box 2.1, *continued*

enhanced risk management. Mobile money adoption can facilitate the transaction of transfers, both remittances and public transfers. These developments can lead to an expansion in per capita consumption in the long term and, potentially, to a reduction in the incidence of extreme poverty and food insecurity.

Doubling the current share of mobile money users in Senegal by 2020—an increase from 31.8 percent to 63.6 percent per the goals set by WAEMU reforms—would have the following welfare implications, according to the simulation exercise. It would raise ownership among women from 29 percent to

57 percent (see figure B2.1.1), induce increased account access for individuals in rural areas from 27 percent to 62 percent, and reduce the mobile money gap between the top and bottom quintiles by 5.1 percentage points. In turn, improved access can potentially increase the use of mobile accounts for sending or receiving payments and private transfers. Indeed, simulation results estimate an increase in remittances (of those potential receivers) by 70 percent, and a doubling of private payments related to labor, such as wages and payments for self-employment and agricultural activities.

FIGURE B2.1.1

Mobile money account ownership, by gender and simulation methods, 2017

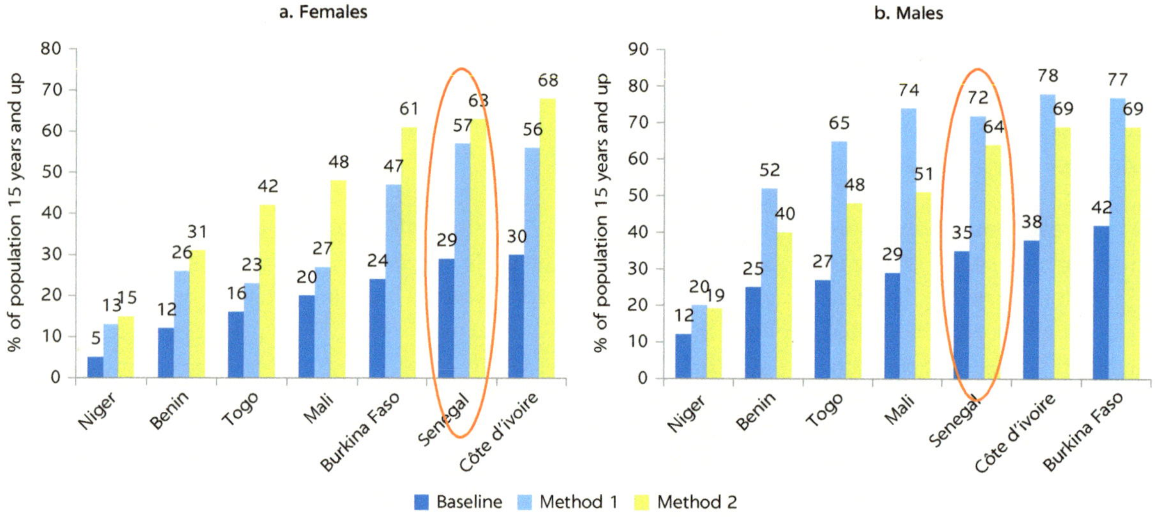

Source: Global Findex 2017 database.
Note: Two methods are employed to simulate the expansion. Method 1 uses a probabilistic model of the likelihood of owning a mobile money account, estimated based on current access to mobile financial services in each country. Method 2 predefines new potential users of mobile money services based on particular conditions faced by individuals, and then applies a random selection among these candidates to fill the quotas set. The figures reported in the text correspond to method 1.

Source: Box 2.1 is based on results presented as part of the Poverty and Social Impact Analysis of the World Bank WAEMU Financial Inclusion Regional Development Policy Operation Project (P171234).

on access to fixed broadband internet—using the presence of terrestrial transmission nodes as a proxy—and 2G coverage maps.[21] The study employs regression analysis to evaluate the association between digital infrastructure and local economic activity.

Expansion of ICT coverage is associated with modest improvements in local economic activity and poverty reduction (figure 2.4). A 1 percentage point increase in 2G connectivity is associated with a 0.055 percentage point decline in

FIGURE 2.4

Effect of 2G coverage on poverty and luminosity in Senegal, 2005 and 2013

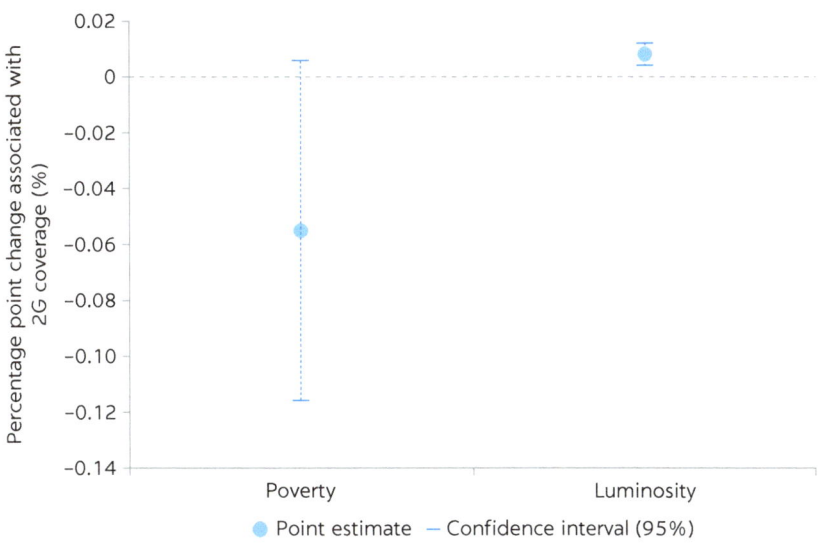

Source: Enamorado et al. 2020.
Note: Point estimates are at 5 percent confidence intervals. Poverty point estimate is the percentage change associated with 2G coverage. The luminosity point estimate represents the mean value change of the nighttime luminosity scale. Ordinary least squares regressions for poverty and luminosity are performed at the commune level.

the poverty rate. There is also evidence of a positive correlation between local economic growth and the expansion of both mobile phone technology and fixed broadband internet. However, these relationships cannot be treated as causal because of (a) data limitations, and (b) the sensitivity of results to how welfare is measured and whether demographic and geographic controls are included in the analysis.

POLICIES TO BROADEN AFFORDABLE INTERNET AVAILABILITY FOR ALL

Having reliable and timely evidence of how mobile broadband could improve welfare is key to informing the policy discussion on DTs. As in other African countries, mobile broadband is the dominant channel through which people access the internet in Senegal. This chapter has highlighted the main findings of recent analyses conducted on the subject in the country—both in what drives and what hinders mobile broadband adoption, as well as what its effects are once it has been adopted. The results show that having purchasing power to acquire technology services is the most important factor for mobile broadband internet adoption. With regard to impact, mobile broadband internet coverage in Senegal is indeed associated with higher household consumption and lower poverty rates. At the local level, expansion in ICT coverage is associated with modest improvements in local economic activity and poverty reduction. Although these analyses are not without limitations, they provide useful conclusions to inform the policy discussion on strategies to enhance the welfare effects of mobile broadband internet access.

Improving the availability of affordable digital infrastructure—particularly in rural areas, including policies geared toward universal 3G coverage—is key to preventing a widening digital divide. None of the benefits of internet access can be realized without being able to adopt the technology first. The analysis of factors of adoption shows that individuals who are poor, illiterate, rural, and female are less likely to adopt mobile internet. Overall, rural areas lag urban ones when it comes to mobile internet adoption, especially due to limited network coverage. This rural/urban divide can have significant economic and welfare implications. Improving the availability of affordable digital infrastructure (box 2.2) can encourage new users and unleash economic opportunities. Policies geared toward universal coverage of 3G mobile service are among the most consequential to reach excluded populations, including people in rural and remote areas who live out of reach of traditional cellular mobile networks. Policies that make headway toward this goal already have the support of the Broadband for All Working Group, which offers an action plan for universal broadband connectivity in Africa.[22] Nevertheless, the feasibility of these efforts will require the continued participation from the government, the private sector, and development partners.

Given the importance of purchasing power as a determinant of broadband mobile internet adoption, the chapter has called attention to the need for tools to reduce budget constraints as well as address other barriers. Policy instruments like social assistance direct transfers can help ease budget constraints for the most income-constrained households and, in turn, facilitate more technology adoption. According to Senegal's *Systematic Country Diagnostic*, ample scope exists for increasing the poverty-reducing effect of direct transfers in the country; some means are increasing expenditures, improving the quality and efficiency of spending, and further enhancing targeting (World Bank 2018). Another important tool to ease budget constraints is to continue to remove barriers—including illiteracy and lack of collateral—to financial inclusion. Another finding from the study suggests that networks—that is, the number of friends using messaging services—are an important driver of adoption. Encouraging a zero price for messaging applications in a low-usage package (only attractive to the lowest-income households) can also incentivize adoption. At the same time, reducing budget constraints for the poor is crucial but not sufficient to improve their access to mobile internet services. Policy design needs to consider the fact that adoption does not directly translate into usage for everyone, whereby specific attention must be given to vulnerable groups. In addition, given the importance of literacy and other skills, continuing investments in education—though they will take time to materialize—can help increase the adoption rate of mobile internet services.

Increased competition in digital infrastructure—such as through more intense rivalry between a higher number of mobile operators and a reduction in the market power of the dominant operator—has measurable benefits on welfare by reducing prices and incentivizing new entrants. Box 2.3 shows the results of an exercise simulating the distributional effects of a rise in competition in the mobile internet market in Senegal through its effects on the price of services. Market concentration (and high prices) in the sector affect all households. The simulation results show that increasing competition—from three to seven firms—leads to a 31 percent reduction in prices, which increases households' purchasing power and leads to welfare gains. Specifically, the rise in competition in the mobile internet market could lead to a reduction in poverty in the

Enhancing digital infrastructure policies and regional coordination

Government policies to deepen affordable digital infrastructure include a range of measures (see Foch 2019 and World Bank 2020). One measure is the implementation of Senegal's recently adopted infrastructure sharing policy (with adequate safeguards to prevent anticompetitive conduct). New evidence suggests that a shared rural network could particularly generate high cost savings (figure B2.2.1). Next steps could include developing effective regulations to implement the secondary legislation on infrastructure sharing and preparing a national fiber-optic plan.

Improving the management of public telecommunications infrastructure by involving the private sector through a public-private partnership (PPP) could also be valuable. The objective would be to pool the national public infrastructure under a patrimonial company and entrust its management to a competent private operator under a PPP arrangement (see Foch 2019 and World Bank 2020). Next steps could include finalizing the concessioning of the State Informatics Agency (Agence de l'Informatique de l'État, ADIE) network and assessing the benefits from more

effective management of other public digital assets such as those deployed by the Senegal National Electricity Agency (SENELEC) or being deployed by the Senegalese national highways and railways network.

A diagnosis of existing infrastructure in Senegal shows an urgent need to provide private operators with passive infrastructure so that they can extend their network coverage at low cost to unserved areas (through network-extension, passive infrastructure such as fiber-optic backbones and shared pylons). The service of areas eligible for universal service could be achieved through a patrimonial approach, which consists of deploying public infrastructure to cover areas that are not profitable for the provision of the required services. This type of project would involve the deployment of broadband infrastructure open to all electronic communications operators. The operation of this infrastructure would be entrusted to a private wholesale operator of infrastructure that would market its services to retail operators. The state can also choose to use the "pay or play" mechanism allowing operators to reduce their contribution to the Universal Services Fund (Fonds de Développement du Service Universel des Télécommunications, FDSUT) up to the net cost of universal access/service projects they agree to implement. It will be important to finalize the FDSUT consultations with private players and launch calls for tenders to invest in new digital infrastructure projects.

Senegal could also take a range of steps to promote regional benefits through increased regional harmonization. One useful direction is to help ensure coordination at the supranational level regarding the WAEMU electronic communications framework and Economic Community of West African States (ECOWAS) rules, promoting closer coordination between national and regional sectoral regulators as well as competition agencies. Among others, this coordination could include fully implementing the ECOWAS roaming regulations. On regional data markets, Senegal could take the lead in advocating for the elimination of restrictive (or lack of) rules across ECOWAS countries to spur development of fully regional data markets. Regarding security, an ECOWAS directive provides a list of offenses related to ICTs, compelling member states to adapt their

FIGURE B2.2.1

Effect of shared digital infrastructure on savings

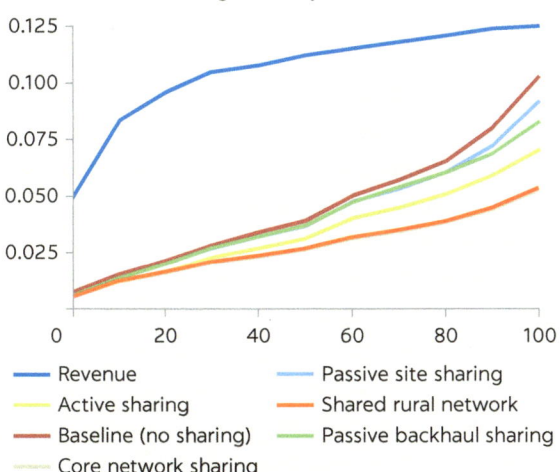

Source: Oughton 2020. Results obtained from a cost model under an assumed average revenue per user of US$2.00–US$8.00 depending on population density, around 60 percent coverage per geographic area, and target data speed of 2 Mbps.

continued

Box 2.2, *continued*

procedural and criminal laws to address cybercrime issues, and promotes international cooperation on cybersecurity. Although implementation is required by all member states, many either have no relevant legislation or are still in the process of adopting it. Senegal stands out as the only country that introduced cybercrime legislation before the ECOWAS directive (World Bank 2021).

More work is needed to explore the extent of benefits, feasibility, and best implementation modalities of deeper regional market integration that can create bigger markets with more competition and reduced costs—supported by regional spectrum auctions to allow for the entry of new regional-level operators and cross-border trading of spectrum between existing operators. An enhanced regional approach to regulation, including the widening of national markets into a fully regional market, is likely to generate the following short-term cost efficiencies, in addition to longer term competition and innovation-related benefits: (a) cost savings as a product of linking sites to the nearest nodes rather than to those located within national borders but that are more distant; (b) the reduction of site administration costs, particularly by allowing the grouping of existing national companies that are part of the same regional operators to serve several countries through one corporate entity and interact with only one regulatory body; (c) the reduction of spectrum acquisition costs; and (d) economies of scale, including by purchasing equipment in larger quantities. For the case of a 4G network, the cost savings have been estimated to be as high as 20 percent, though these findings are preliminary and depend on assumptions regarding cost reductions (Oughton 2020).

medium-to-long run of 0.67 and 0.85 percentage points, respectively. The price reduction also increases new customers (by 2.9 percentage points), with the largest rise in adoption in the third and fourth quintiles of the distribution. These findings highlight the importance of deepening the ongoing reforms in Senegal to increase competition in mobile broadband services. Policy efforts to broaden affordable internet access for all could include promoting a sustainable presence of the three existing internet service providers (ISPs) in the market and facilitating further entry of new wholesale and retail players. Among other measures, this ISP action could help address rents in the mobile oligopoly, including the dominant position of Sonatel-Orange across the international access, mobile, and backbone infrastructure—as well as addressing how this dominance extends to mobile money and data ownership, and to e-commerce (through indirect ownership in Jumia).

The regulatory aspects mentioned draw attention to the importance of regulatory frameworks in the way DTs contribute to shaping the inclusiveness of digital-driven growth. A detailed regulatory analysis is beyond the scope of this overview. Yet some of the regulatory dimensions touched on, including digital infrastructure policies (and the value of shared rural networks) and the potential welfare effects of a rise in telecom sector competition (which makes the case for reforms to lift barriers to entry for providers and operators), emphasize their importance. Regulatory frameworks can play a significant role in the expansion of DTs, such as in relation to promoting new entrants (providers, operators, firms, and customers) and innovation. They are also key to supporting digital business models and to addressing issues around privacy and data sharing. The *World Development Report 2021: Data for Better Lives* will help shed further light on these issues.

BOX 2.3

Distributional effects of competition in Senegal's telecommunications market: A microsimulation approach

Competitive and efficient environments are crucial to fully reaping the benefits of ICTs for welfare. The lack of competition across ICT markets limits accessibility, affordability, and quality of services, while hindering private sector investment and innovation. Recent simulation exercises show that increased competition in digital infrastructure in Mexico and Djibouti can significantly improve welfare by reducing the price of ICT services (Rodríguez-Castelán et al. 2019; Decoster et al. 2019). The importance of competition for DT adoption and use has been stressed by other studies, including competition's role as a potential determinant of the large disparities in cell phone coverage systems in Sub-Saharan Africa (Buys et al. 2009; Howard and Mazaheri 2009).

The Senegalese telecommunications market is often categorized as having low competition, resulting in low quality and high prices. Against this backdrop and with the objective of driving the digital transformation of the economy, the government has implemented a series of reforms to overcome barriers to entry for new internet service providers and wholesale infrastructure operators in the country.

Using the Welfare and Competition tool (WELCOM) developed by the World Bank, this exercise simulates the distributional effects of a rise in competition in the mobile internet market in Senegal and its effects on the price of services.[a] Market concentration—and high prices—in this sector affect all households. As shown in figure B2.3.1, the poor spend 1.0 percent of their total expenditures on mobile internet services, while the expenditures of the richest quintile lie at 4.0 percent. The simulation relies on the assumption that the mobile internet market behaves as an oligopoly. For the purpose of this exercise, the scenario assumes increasing competition from three to seven firms in the mobile internet market and a price elasticity of demand of –1.5.[b] The data used are obtained from Senegal's EHCVM 2018/2019 survey.

FIGURE B2.3.1

Average expenditure share of 3G services

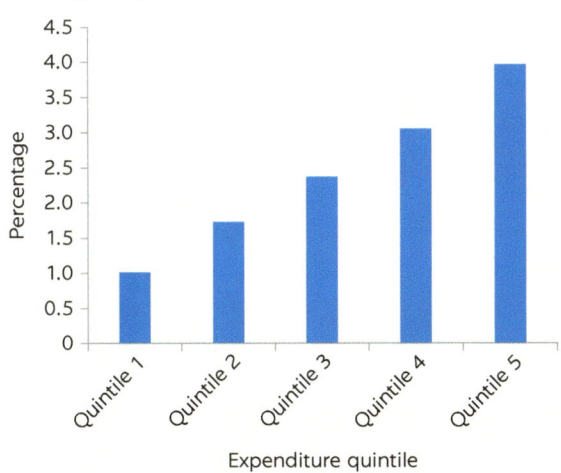

FIGURE B2.3.2

Change in poverty as a result of mobile internet firms' entrance

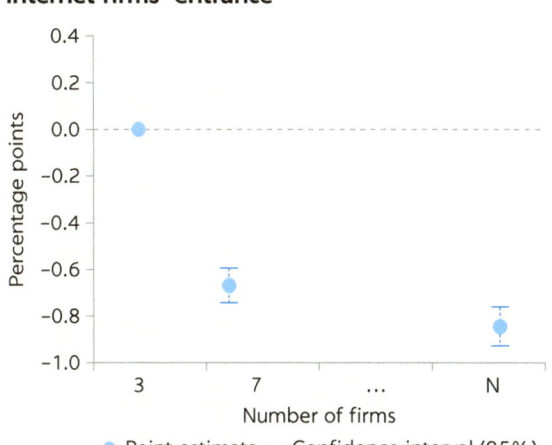

Source: Based on EHCVM 2018/2019.
Note: Households with positive expenditures. Simulation assumes that mobile broadband market behaves as an oligopoly. The scenario assumes increasing competition from three to seven firms and a price elasticity of demand of –1.5, resulting in a 31 percent price reduction. Mobile internet adoption and expenditures are considered for individuals 15 years and older.

continued

Box 2.3, *continued*

The simulation results show that increasing competition could achieve a lower-bound impact on poverty reduction in the medium-to-long term of 0.67 percentage points (figure B2.3.2). This number would mean lifting about 106,000 people above the poverty line. In the long run, the estimates suggest a poverty reduction of 0.85 percentage points. The mechanism behind this result is that increased competition would lead to lower prices, specifically to a 31 percent reduction. As households are able to consume services at lower prices, their purchasing power rises, leading to gains in welfare.

All quintiles would benefit from higher competition in mobile internet, particularly the top one, because higher disposable income—resulting from the price reduction—allows individuals to spend more on mobile internet (figure B2.3.3).[c] The large relative welfare incidence in the top quintile is also explained by its high degree of 3G coverage, at 98.2 percent (compared with 85.7 percent for the poorest quintile; this lower coverage reduces the margin of adoption and

welfare impact for the poorest households). Considering current users alone (not taking into account new users resulting from the price reduction), higher competition would induce a total relative welfare gain of 1.74 percent.

Welfare gains from increased competition could also be significant because of the additional technology adoption by new users who were previously priced out. Using the simulated (lower) prices resulting from higher competition, it is possible to estimate the uptake of users and calculate an expected change in the quantity consumed. Estimates suggest that a 31 percent reduction in prices as a result of higher competition leads to a 2.9 percentage point increase in the number of users, where the third and fourth quintiles see the largest relative increase in adoption, benefiting middle-income households relatively more (figure B2.3.4). This change represents 258,000 new entrants in the market, increasing the total coverage of mobile internet from 28.4 to 31.3 percent of the population.[d] New market entrants would subsequently lead

FIGURE B2.3.3

Relative effect on households' budget associated with higher competition in 3G services[a]

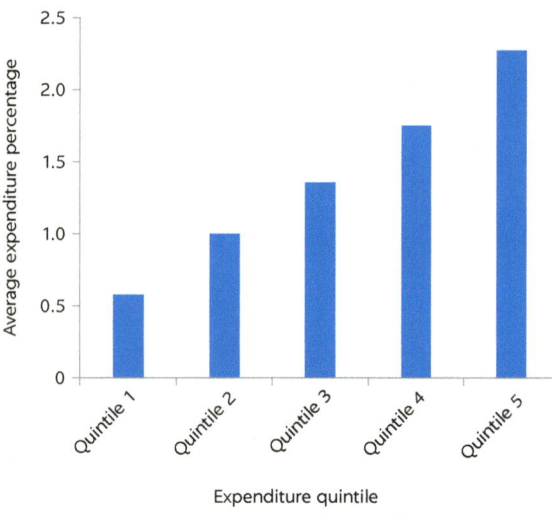

FIGURE B2.3.4

Relative effect on households' budget and welfare monetary incidence associated with higher competition in 3G services[b]

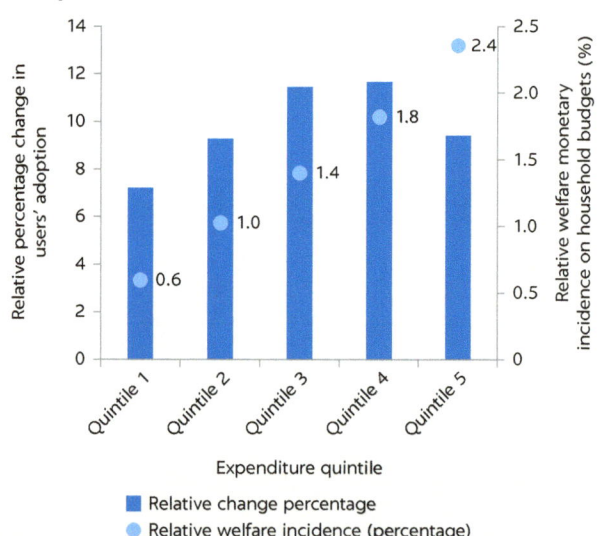

Source: Simulations based on EHCVM 2018/2019.
Note: Simulation relies on the assumption that mobile broadband market behaves as an oligopoly. Simulation scenario assumes increasing competition from three to seven firms and a price elasticity of demand of −1.5, resulting in a 31 percent price reduction.
a. Based on current users among individuals above 15 years of age by expenditure quintile.
b. Based on all consumers (= current + new users) among individuals above 15 years of age by expenditure quintile.

continued

Box 2.3, *continued*

to a welfare increase of 0.05 percent in average per capita household expenditures, resulting in a 1.79 percent total relative welfare gain. The welfare effect is dominated by current users: between 200,000 and 300,000 existing users around the poverty line would perceive a welfare increase, compared with 30,000 new users. When the welfare effect is adjusted for new entrants only, it is still of a considerable magnitude (a 2.8 percent increase in total household per capita expenditures for the poorest quintile and a 1.8 percent increase for the richest). The combined welfare effect of competition on both current and new users would induce total poverty reduction of 0.72 percentage points.

a. The WELCOM (Welfare and Competition) simulation tool was developed by the Poverty and Equity GP at the World Bank. For more information, see Rodríguez-Castelán et al. (2019), http://dasp.ecn.ulaval.ca/webwel/welcom.html. Regarding data, the latest wave of Senegal's household survey (WAEMU) does not cover expenditures on mobile internet. To the best of our knowledge, there are no other indicators available at a granular level to measure expenditures on mobile internet. Therefore, to determine household expenditures on mobile internet, we create a proxy that identifies individuals with positive expenditures on prepaid mobile phone cards and airtime/data transfers, who reported accessing the internet through their mobile phones. This proxy can be considered an upper bound for potential mobile internet users, where the underlying assumption is that all users who claim to access the internet through their mobile phones (and have positive expenditures on mobile services) own smartphones that enable them to access the internet.
b. There is no price elasticity of demand (PED) specific to the Senegal mobile services market available. The estimate used should be taken with caution as it is from 2005 and corresponds to values obtained for a pool of developing countries (Waverman, Meschi, and Fuss 2005). However, given the scant empirical evidence on PEDs of mobile services in developing countries, particularly in SSA, using this estimate appears to be a reasonable solution.
c. Relative welfare is the per capita welfare monetary incidence divided by total per capita expenditures.
d. The 28.4 percent coverage serves as the baseline adoption threshold for this simulation as it considers individuals who reported accessing the internet through their mobile phones and had positive expenditures on mobile services. This figure differs from the 34.2 percent coverage mentioned in previous sections of the document, which does not consider expenditures.

Improving digital literacy, local linguistic content, and the quality of digital services can help enhance the positive effects of mobile broadband internet access for users. The evidence presented in this chapter showcases the concrete benefits of mobile broadband for household welfare, in increased consumption and lower poverty reduction. Digital literacy is an important foundation of adopting and using digital technologies, highlighting the importance of continuing to invest in education. Another important finding from the analysis is that the high degree of coverage has not necessarily translated into usage, for several reasons. A rural/urban divide prevails, highlighting a quality gap between areas. Improving the quality of services is key to bridging this divide. Another area for policy refers to the promotion of local linguistic content. If Wolof is the main preferred spoken language and it is not, in fact, widely available online, then people who speak Wolof and not French would find that to be an obstacle to adoption and use. Future work can focus on a better understanding of the role of accessible content, as well as on the uses of mobile internet by households, such as in leisure versus productive use—for example, for job searches.

Promoting financial inclusion and the productive use of digital technologies—including mobile money, digital payments, and electronic commerce—can improve people's livelihoods. As described in box 2.1, expanding digital financial services can increase the delivery and affordability of financial services to populations underserved by the traditional financial system. The analysis shows that expanding access to mobile money accounts in Senegal can reduce differences in account ownership by gender, education, rural/urban area, income, and labor force participation status, reducing horizontal inequality. Expanding the use of mobile accounts will likely lead to an increase in remittances and private payments. These findings highlight the importance of mobile money and of policies that foster interconnectivity between banks, reduce transaction costs, and increase the supply of mobile money agents in rural areas. Subsequent research can shed light on the implications of mobile broadband as a cornerstone for private and public platforms, including mobile money and e-government applications, which so far tend to be based mostly around 2G technology in the region.

NOTES

1. This chapter is based on the results and analysis from the following background papers: Atiyas and Doğanoğlu (2020); Enamorado et al. (2020); Masaki, Granguillhome Ochoa, and Rodríguez-Castelán (2020); and Rodríguez-Castelán et al. (2021a, 2021b). The policy discussion incorporates inputs provided by Izak Atiyas. The chapter benefited from comments from Mark Dutz.
2. The literature identifies socioeconomic characteristics such as education, gender, income, age, digital literacy, or household size as drivers related to the adoption and use of the internet (Coelho, Silva, and Ehrl 2019; Goldfarb and Prince 2008; Grazzi and Vergara 2014; Kongaut and Bohlin 2016; Martínez-Domínguez and Mora-Rivera 2020; and Nishijima, Ivanauskas, and Mori Sarti 2017). Existing studies from SSA reach similar findings (Penard et al. 2012, 2015; Birba and Diagne 2012; Gillwald, Milek, and Stork 2010). Most of this research, however, concentrates on fixed rather than mobile broadband. This distinction is important in the context of African countries, where most people access the internet through mobile phones rather than through fixed broadband (ITU 2019). Country-level characteristics can also play a role on internet coverage and adoption. Electricity access appears to be a key driver of internet adoption in poor countries (Armey and Hosman 2016). And increased competition among digital service providers can reduce prices and allow new entrants to adopt internet services (Decoster et al. 2019; Rodríguez-Castelán et al. 2019). Indeed, competition appears to have had a positive and significant effect on cell phone coverage systems in SSA (Buys et al. 2009). Other research considers the role of prices and variable and fixed costs as factors behind internet adoption. Studies from the United States show that consumers tend to place a premium on unlimited data and unmetered pricing, and a lower value on bandwidth (Liu, Prince, and Wallsten 2018; Varian 2002). Others focus on the surplus generated from residential broadband usage and the degree of capture of internet service providers (Nevo, Turner, and Williams 2016). A cross-country OECD study finds that the demand for internet services is price-inelastic—where people continue demanding it even as its price increases—while its income elasticity is greater than one, suggesting that relative to changes in income, the internet is more of a luxury good (Goel et al. 2006).
3. This section draws results from the background paper "Mobile Internet Adoption in West Africa" (Rodríguez-Castelán et al. 2021).
4. The number of active mobile broadband subscriptions per 100 inhabitants in Africa in 2019 was 34, compared with 0.4 for fixed broadband subscriptions (ITU 2019).
5. These variables are defined at the household level. Cell phone age corresponds to the question "For how long have you owned the following item?"
6. All variables pertain to characteristics of individuals above 15 years of age. Each tabulation is carried out at the individual level adjusting for household weights. Connectivity should thus be interpreted as the percentage of individuals who access the internet through their mobile phone.
7. The results presented in this paragraph and the next one regarding the difference of means between connected and not-connected individuals compared with sociodemographic characteristics are all significant at the 1 percent level, except where marked otherwise.
8. The analysis models adoption as a two-stage process, implementing a Heckman-corrected probit model to account for selection bias. It is assumed that there is a fundamental relationship between households that are covered by 3G (first stage) and those that decide to adopt mobile internet (second stage). In addition to a welfare metric (household expenditures), the analysis incorporates a human capital dimension (education, sector of employment, language), access to different internet modalities and household assets (computer,

television, and tablet ownership), and a price proxy for mobile internet. The analysis also controls for selection bias by truncating the sample conditional to those households with 3G coverage. Coverage data were obtained directly from Collins Bartholomew, a digital mapping provider, and the three major mobile operators in Senegal: Expresso, Orange (or its local subsidiary, Sonatel), and Tigo. Mobile internet price is obtained by calculating the median expenditure of prepaid mobile phone cards and airtime/data transfers among mobile internet users in each country's geographic area for which the survey is representative. This value is then computed as a share of total consumption at the same geographic level to adjust for cost of living. This value is then imputed to each individual observed in the microdata.

9. Interpretations of the model are based on average marginal effects (AME). The results presented here differ slightly from previous estimates reported by Zeufack et al. (2020) as the model now controls for 3G coverage by implementing a Heckman-corrected probit model (though the parsimonious model yields similar results) and incorporates the price proxy derived from individual-level expenditures on prepaid mobile phone cards and airtime/data transfers.

10. This change is equivalent to a one unit increase in the standard deviation of the log of per capita expenditures. Currency conversion based on an exchange rate of US$1.00 = CFAF 555.45 in 2018. International Financial Statistics (database), International Monetary Fund, Washington, DC, https://data.imf.org/?sk=4c514d48-b6ba-49ed-8ab9-52b0c1a0179b.

11. This change is equivalent to a 1-unit increase in the standard deviation of the mobile data price proxy derived from individual-level expenditures on prepaid mobile phone cards and airtime/data transfers.

12. This section draws results from the background paper "Internet Adoption in Senegal," Atiyas and Doğanoğlu (2020).

13. This section draws results from the background paper "Broadband Internet and Household Welfare in Senegal" (Masaki, Granguillhome Ochoa, and Rodríguez-Castelán 2020).

14. Data on terrestrial backbone networks were obtained from http://www.africaband widthmaps.com. Internet traffic in countries runs first through national "backbone" or fiber-optic networks, which are then connected to end users through last-mile infrastructure such as fiber cables, copper cables, wireless transmission, or cell phone towers among others. See Hjort and Poulsen (2019) for a detailed discussion.

15. Results are robust after controlling for household demographics and other spatial characteristics (for example, region-fixed effects, road density, nighttime lights, or elevation) as well as for access to complementary digital infrastructures, such as 2G coverage or fixed broadband internet. Results are also robust to an IV approach, using distance to 3G coverage in neighboring areas as an instrument.

16. Taking advantage of longitudinal data in Nigeria, Bahia et al. (2020) prove that the parallel-trends hypothesis holds for their difference-in-difference analysis. In contrast, the results for Senegal likely overestimate the real effect of internet access, because the digital infrastructure is almost surely located in more affluent areas with lower poverty rates and higher household consumption.

17. Employment is defined as working-age (15–64 years) individuals who worked at least one hour in the past seven days. Salaried/wage employment includes those employees who work in a place that is not their own farm or in a business not run by their own household.

18. Socioeconomic characteristics—such as age, education, and income—appear to influence the adoption of DTs for different uses. A study using data from Cameroon finds that young people tend to use the internet for leisure purposes while users who are older, more educated, and computer savvy are more likely to use it to search for information (Penard et al. 2015). Analogously, a study of rural users in Mexico finds that young people are more likely to use the internet for entertainment, while working-age people go online for information, communication, and e-commerce activities (Martínez-Domínguez and Mora-Rivera 2020). The study also finds that higher education increases the types of uses. In the United States, Goldfarb and Prince (2008) show that low-income people are more likely to do time-consuming, inexpensive activities online. However, the authors find that more hours using the internet are related to an increase in the use of more "valuable" activities, such as for e-government, researching purchases, telemedicine, and news.

19. The literature suggests that the internet may have heterogeneous effects by location, benefiting some geographic areas over others, as its effect in reducing frictions and costs varies

according to local factors (Greenstein, Forman, and Goldfarb 2018). But little is known about the potential impact of DTs on regional poverty trends, particularly for developing countries. Most evidence on the territorial digital divide concentrates on developed economies. A study from the United States finds that broadband increased employment rates more in low-density rural areas than in urban ones (Kolko 2012). Conversely, Forman, Goldfarb, and Greenstein (2012) find that the internet exacerbates regional income inequality, based on data from US counties. As for developing countries, research from Brazil suggests that, while positive, the incidence of broadband on productivity is not uniform across regions, with evidence of regional convergence (Jung and López-Bazo 2020). Faster download speed and critical mass, which account for network externalities, it is argued, tend to enhance the economic impact of broadband.

20. This section draws from the background paper titled "Local Welfare Effects of Digital Technologies in Senegal" (Enamorado et al. 2020).

21. As in the previous analyses, 2G coverage maps are obtained from Collins Bartholomew, while African Bandwidth Maps is the source for maps of the fiber-optic networks.

22. The plan calls for the following: Ensuring that the commercial broadband market is open and structurally prepared for competitive private investment; providing public/donor funding support for larger, high-cost infrastructure investments to reduce risk and increase commercial viability; providing direct funding support for extending affordable broadband access to commercially challenging rural and remote areas, and to women and low-income users under a "Mobilizing Finance for Development" approach; reducing noneconomic costs and risks of market entry and investment; expanding the market through government procurement and implementation of broadband-based digital services, networks, and facilities; and ensuring that the technical skills needed to operate and maintain digital infrastructure are increasingly available in the region. For more, see *Connecting Africa through Broadband: A Strategy for Doubling Connectivity by 2021 and Reaching Universal Access by 2030* (World Bank 2019a).

REFERENCES

Armey, Laura E., and Laura Hosman. 2016. "The Centrality of Electricity to ICT Use in Low-Income Countries." *Telecommunications Policy* 40 (7): 617–27.

ARTP (Autorité de Régulation des Télécommunications et des Postes). 2019. Observatoire de L'Internet. "Tableau de bord a au 31 decembre 2019."

Atiyas, Izak, and Toker Doğanoğlu. 2020. "Using the RIA Data Set to Explore Correlates of Mobile Internet Use in Senegal." Internal report, World Bank, Washington, DC.

Bahia, Kalvin, Pau Castells, Genaro Cruz, Takaaki Masaki, Xavier Pedrós, Tobias Pfutze, Carlos Rodríguez-Castelán, and Hernan Winkler. 2020. "The Welfare Effects of Mobile Broadband Internet: Evidence from Nigeria." Policy Research Working Paper 9230, World Bank, Washington, DC.

Birba, Ousmane, and Abdoulaye Diagne. 2012. "Determinants of Adoption of Internet in Africa: Case of 17 Sub-Saharan Countries." *Structural Change and Economic Dynamics* 23 (4): 463–72.

Buys, Piet, Susmita Dasgupta, Tim Thomas, and David Wheeler. 2009. "Determinants of a Digital Divide in Sub-Saharan Africa: A Spatial Econometric Analysis of Cell Phone Coverage." *World Development* 37 (9): 1494–505.

Calderón, César, Gerard Kambou, Vidjan Korman, Megumi Kubota, and Catalina Cantú. 2019. "An Analysis of Issues Shaping Africa's Economic Future." *Africa's Pulse*, no. 19 (April), World Bank, Washington, DC.

Chun, N., and H. Tang. 2018. "Do Information and Communication Technologies Empower Female Workers? Firm-Level Evidence from Viet Nam." Asian Development Bank Institute Working Paper 545, Tokyo.

Coelho, Florângela, Thiago Silva, and Philipp Ehrl. 2019. "Internet Access in Brazilian Households: Evaluating the Effect of an Economic Recession." In *New Knowledge in Information Systems and Technologies*, edited by Álvaro Rocha, Hojjat Adeli, Luís Paulo Reis, and Sandra Costanzo, 716–25. *Advances in Intelligent Systems and Computing* book series, vol. 931. Cham, Switzerland: Springer.

Decoster, Xavier, Gabriel Lara Ibarra, Vibhuti Mendiratta, and Marco Santacroce. 2019. "Welfare Effects of Introducing Competition in the Telecom Sector in Djibouti." Policy Research Working Paper 8850, World Bank, Washington, DC.

De los Rios, Carlos. 2010. "Welfare Impact of Internet Use on Peruvian Households." Instituto de Estudios Peruanos, Lima.

EHCVM (Enquête Harmonisée sur les Conditions de Vie des Ménages 2018/2019). Agence Nationale de la Statistique et de la Demographie.

Elbers, Chris, Jean O. Lanjouw, and Peter Lanjouw. 2003. "Micro-Level Estimation of Poverty and Inequality." *Econometrica* 71 (1): 335–64.

Enamorado, Ted, Takaaki Masaki, Carlos Rodríguez-Castelán, and Hernan Winkler. 2020. "Local Welfare Effects of Digital Technologies in Senegal." Internal report, World Bank, Washington, DC.

Fernandes, Ana, M. Aaditya Mattoo, Huy Nguyen, and Marc Schiffbauer. 2019. "The Internet and Chinese Exports in the Pre–Ali Baba Era." *Journal of Development Economics* 138: 57–76.

Foch, A. 2019. "Accélérer le programme de réformes en matière d'infrastructures et de services haut débit pour promouvoir l'essor de l'économie numérique." In *Sénégal: Notes de Politiques Economiques et Sociales,* 92–117. Washington, DC: World Bank.

Forman, Chris, Avi Goldfarb, and Shane Greenstein. 2012. "The Internet and Local Wages: A Puzzle." *American Economic Review* 102 (1): 556–75.

Gillwald, Alison, Anne Milek, and Christoph Stork. 2010." Gender Assessment of ICT Access and Usage in Africa." *Towards Evidence-Based ICT Policy and Regulation,* vol. 1, Policy Paper 5, Research ICT Africa.

Goel, Rajeev, Edward Hsieh, Michael Nelson, and Rati Ram. 2006. "Demand Elasticities for Internet Services." *Applied Economics* 38 (9): 975–80.

Goldfarb, Avi, and Jeffrey Prince. 2008. "Internet Adoption and Usage Patterns Are Different: Implications for the Digital Divide." *Information Economics and Policy* 20: 2–15.

Goyal, Aparajita. 2010. "Information, Direct Access to Farmers, and Rural Market Performance in Central India." *American Economic Journal: Applied Economics* 2 (3): 22–45.

Grazzi, Matteo, and Sebastian Vergara. 2014. "Internet in Latin America: Who Uses It? . . . and for What?" *Economics of Innovation and New Technology* 23 (4): 327–52.

Greenstein, Shane, Chris Forman, and Avi Goldfarb. 2018. "How Geography Shapes—and Is Shaped by—the Internet." In the *New Oxford Handbook of Economic Geography,* edited by Gordon Clark, Maryann Feldman, Meric Gertler, and Dariusz Wójcik. New York: Oxford University Press.

Hasbi, Maude, and Antoine Dubus. 2019. "Determinants of Mobile Broadband Use in Developing Economies: Evidence from Sub-Saharan Africa." Working Paper hal-02264651, HAL.

Hjort, Jonas, and Jonas Poulsen. 2019. "The Arrival of Fast Internet and Employment in Africa." *American Economic Review* 109 (3): 1032–79.

Howard, Philip N., and Nimah Mazaheri. 2009. "Telecommunications Reform, Internet Use and Mobile Phone Adoption in the Developing World." *World Development* 37 (7): 1159–69.

ITU (International Telecommunications Union). 2019. *Measuring Digital Development: Facts and Figures 2019.* Geneva: ITU.

ITU (International Telecommunications Union). 2020. *Measuring Digital Development ICT Price Trends 2019.* Geneva: ITU.

Jung, Juan, and Enrique López-Bazo. 2020. "On the Regional Impact of Broadband on Productivity: The Case of Brazil." *Telecommunications Policy* 44 (1, 101826).

Kaila, Heidi, and Finn Tarp. 2019. "Can the Internet Improve Agricultural Production? Evidence from Viet Nam." *Agricultural Economics* 50 (6): 675–91.

Kolko, Jed. 2012. "Broadband and Local Growth." *Journal of Urban Economics* 71 (1): 100–113.

Kongaut, Chatchai, and Erik Bohlin. 2016. "Investigating Mobile Broadband Adoption and Usage: A Case of Smartphones in Sweden." *Telematics and Informatics* 33 (3): 742–52.

Liu, Yu-Hsin, Jeffrey Prince, and Scott Wallsten. 2018. "Distinguishing Bandwidth and Latency in Households' Willingness-to-Pay for Broadband Internet Speed." *Information Economics and Policy* 45: 1–15.

Marandino, Joaquin, and Phanindra Wunnava. 2014. "The Effect of Access to Information and Communication Technology on Household Labor Income: Evidence from One Laptop per Child in Uruguay." IZA Discussion Paper 8415, Institute for the Study of Labor, Bonn.

Martínez-Domínguez, Marlen, and Jorge Mora-Rivera. 2020. "Internet Adoption and Usage Patterns in Rural Mexico." *Technology in Society* 60, 101226.

Masaki, Takaaki, Rogelio Granguillhome Ochoa, and Carlos Rodríguez-Castelán. 2020. "Broadband Internet and Household Welfare in Senegal." Policy Research Working Paper 9386, World Bank, Washington, DC.

Menon, Nidhiya. 2011. "Got Technology? The Impact of Computers and Cell Phones on Productivity in a Difficult Business Climate: Evidence from Firms with Female Owners in Kenya." IZA Discussion Paper 5419, Institute of Labor Economics (IZA), Bonn.

Nevo, Aviv, John L. Turner, and Jonathan W. Williams. 2016. "Usage-Based Pricing and Demand for Residential Broadband." *Econometrica* 84 (2): 411–43.

Nishijima, Marislei, Terry Macedo Ivanauskas, and Flavia Mori Sarti. 2017. "Evolution and Determinants of Digital Divide in Brazil (2005–2013)." *Telecommunications Policy* 41 (1): 12–24.

Oughton, Edward. 2020. "Policy Options for Affordable Digital Infrastructure Expansion: A Simulation Model for National and Regional Markets in Africa." Internal report, World Bank, Washington, DC.

Paunov, Caroline, and Valentina Rollo. 2015. "Overcoming Obstacles: The Internet's Contribution to Firm Development." *World Bank Economic Review* 29 (suppl. 1): S192–S204.

Penard, Thierry, Nicolas Poussing, Gabriel Zomo Yebe, and Philemon Nsi Ella. 2012. "Comparing the Determinants of Internet and Cell Phone Use in Africa: Evidence from Gabon." *Communications and Strategies* 86 (2nd quarter): 65–83.

Penard, Thierry, Nicolas Poussing, Blaise Mukoko, and Georges Bertrand Tamokwe Piaptie. 2015. "Internet Adoption and Usage Patterns in Africa: Evidence from Cameroon." *Technology in Society* 42: 71–80.

Ritter, Patricia, and Maria Guerrero. 2014. "The Effect of the Internet and Cell Phones on Employment and Agricultural Production in Rural Villages in Peru." Working paper, University of Piura, Piura, Peru.

Rodríguez-Castelán, Carlos, Abdelkrim Araar, Eduardo A. Malasquez, Sergio Olivieri, and Tara Vishwanath. 2019. "Distributional Effects of Competition: A Simulation Approach." Policy Research Working Paper 8838, World Bank, Washington, DC.

Rodríguez-Castelán, Carlos, Rogelio Granguillhome Ochoa, Samantha Lach, and Takaaki Masaki. 2021a. "Mobile Internet Adoption in West Africa." Policy Research Working Paper 9560, World Bank, Washington, DC.

Rodríguez-Castelán, Carlos, Samantha Lach, Takaaki Masaki, and Rogelio Granguillhome Ochoa. 2021b. "How Do Digital Technologies Affect Household Welfare in Developing Countries? Evidence from Senegal." Policy Research Working Paper 9576, World Bank, Washington, DC.

Salas Garcia, Vania B., and Qin Fan. 2015. "Information Access and Smallholder Farmers' Selling Decisions in Peru." Paper presented at the Agricultural and Applied Economics Association (AAEA) & WAEA Joint Annual Meeting, paper no. 205380, July 26–28, San Francisco.

Varian, Hal R. 2002. "The Demand for Bandwidth: Evidence from the INDEX Project." In *Broadband: Should We Regulate High-Speed Internet Access?*, edited by Robert W. Crandall and James H. Alleman, 39–56. Washington, DC: AEI-Brookings Joint Center for Regulatory Studies.

Viollaz, Mariana, and Hernan Winkler. 2020. "Does the Internet Reduce Gender Gaps? The Case of Jordan." Policy Research Working Paper 9183, World Bank, Washington, DC.

Waverman, Leonard, Meloria Meschi, and Melvyn Fuss. 2005. "The Impact of Telecoms on Economic Growth in Developing Countries." *Vodafone Policy Paper Series* 2: 10–24.

World Bank. 2016. *World Development Report 2016: Digital Dividends*. Washington, DC: World Bank.

World Bank. 2018. *Senegal: Systematic Country Diagnostic*. Washington, DC: World Bank.

World Bank. 2019a. *Connecting Africa through Broadband: A Strategy for Doubling Connectivity by 2021 and Reaching Universal Access by 2030*. Washington, DC: World Bank.

World Bank. 2019b. "Poverty and Equity GP." Technical paper, World Bank, Washington, DC.

World Bank. 2020. "Advisory Services and Analytics on Digital Connectivity and Transformation in Senegal" (P169007). Technical reports prepared for the government of Senegal by the World Bank Digital Development Practice Group, Washington, DC.

World Bank. 2021. *World Development Report 2021: Data for Better Lives*. Washington, DC: World Bank.

Zeufack, Albert G., César Calderón, Gerard Kambou, Megumi Kubota, Catalina Cantú Canales, and Vijdan Korman. 2020. "An Analysis of Issues Shaping Africa's Economic Future." *Africa's Pulse* vol. 22 (October 2020), World Bank, Washington, DC.

3 Enterprises
INNOVATION FOR BETTER JOBS FOR MORE PEOPLE

SENEGAL NEEDS BETTER AND MORE FIRMS

A pressing challenge for Senegal, with its large population of young workers, is to generate enough good jobs for its growing labor supply.[1] Senegal has a high incidence of early-stage business activity but a low entry rate into the formal sector. The 2016 Recensement Général des Entreprises (RGE), which is the most recent and comprehensive establishment-level census conducted by the Agence Nationale de la Statistique et de la Démographie (ANSD), identifies 407,882 economic units, including formal and informal businesses.[2] The data suggest that fewer than 3 percent operate in the formal sector, roughly 12,000 firms.[3] Most activities are informal and driven by subsistence needs. Those entrepreneurs choose that work largely by necessity and would be better allocated as wage earners in growing firms if such jobs were available. But they are not. However, among micro informal enterprises, there is an intermediate group of potentially productive held-back entrepreneurs who are constrained from entering the modern economy by the risks and costs of building the required bundle of capabilities; they have the potential to join the modern economy and create better jobs if they can benefit from support from other small producers like themselves, more sophisticated buyers, and appropriate government support programs.[4]

The formal sector is small—so creating better jobs for more people needs to include boosting the productivity of informal enterprises in addition to better and more private sector formal jobs. A small share (6 percent) of firms in Senegal has 5 or more employees and is responsible for 48 percent of workers and 81 percent of sales (table 3.1). The share of formal firms with 5 or more employees is even smaller, at 1.3 percent of the total number of firms. These formal firms with 5 or more employees employ 28 percent of workers, account for 77 percent of all sales, and are about 7 times more productive (based on sales per worker) than the average establishment in the RGE database.[5] This large difference in productivity is also observed between formal and informal firms, disregarding their size in number of employees (table 3.1). Improving the productivity of both formal and informal firms, and increasing the number of firms over time in the growing formal group, including both more small and medium enterprises (SMEs) and more large firms, could have a significant effect on the overall productivity, number, and quality of jobs in the country.

TABLE 3.1 Share of firms, workers, sales, and relative productivity, by size

	TOTAL				FORMAL			
SIZE	FIRMS (%)	WORKERS (%)	SALES (%)	PROD.ᵃ	FIRMS (%)	WORKERS (%)	SALES (%)	PROD.ᵃ
Micro (0–4)	93.6	52.0	19.5	1.0	1.2	0.8	3.8	2.5
Small (5–19)	5.6	14.6	20.8	1.1	0.9	3.0	17.5	6.9
Medium (20–99)	0.6	9.4	18.6	1.4	0.3	5.5	18.3	7.0
Large (100+)	0.1	24.0	41.1	2.7	0.1	19.9	41.0	7.5
Nonmicro firms (5 or more employees)	**6.4**	**48.0**	**81.5**	**1.1**	**1.3**	**28.0**	**77.0**	**7.0**

Source: ANSD-RGE.

a. Refers to labor productivity, measured as the ratio between the median value of sales per worker in the respective size group and the median value of sales in the total population. Results suggest that a median formal firm with 5 or more employees is 7 times more productive than an average firm.

Yet even formal firms in Senegal are lagging in technology adoption, and the entry rate of new formal firms is low. New measures of technology adoption at the firm level suggest a large technological gap of firms in Senegal compared with the state of Ceará in Brazil, or Vietnam.[6] The results, which are presented in greater detail in the following sections, suggest that firms in Senegal are still relying mostly on manual procedures to perform many of these tasks, even though there is a lot of heterogeneity between formal and informal firms. Moreover, the entry rate of formal firms, which are usually more capable and productive, has been very low in Senegal. Despite improvements in the number of new registered businesses per thousand working-age population, from 0.29 in 2013 to 0.47 in 2018, this number is still low compared with Côte d'Ivoire (0.74), Zambia (1.1), Kenya (0.9), or South Africa (10). The combination of a large technological gap of existing firms and a low entry rate of firms with higher potential leads to a lower likelihood of having the number of firms with capabilities to be more productive, compete domestically and abroad, and scale up, as would be required to generate more jobs with higher earnings.

Senegal needs to improve the quality of existing firms through technological catchup plus increase the entry of better-quality firms. Acting in both dimensions simultaneously is difficult. However, it is critical to speed up the potential of having a larger share of more productive firms that have the larger production and sales needed to absorb more workers with better earnings. Digital technologies (DTs) can play an important role in this process. Evidence for African countries, including Senegal, suggests that the arrival of fast internet increases firm entry, productivity, and exporting, and contributes positively to higher net job creation and income.[7] Yet DTs are not a panacea. Except for mobile phones that have been diffused widely (largely 2G rather than 3G/4G),[8] many firms in Senegal are still not adopting basic DT solutions that ride on the internet, even in places where these technologies are available, such as Dakar. Understanding the reason for this behavior is critical to designing policies that are more effective and complement the current efforts associated with investments in digital infrastructure.[9]

This chapter analyzes the challenges faced by Senegal to generate better and more firms through both within-firm technology upgrading and start-up entrepreneurship dimensions. First, it provides a detailed picture of the current state of technology adoption by firms in Senegal and analyzes the main obstacles they face to increase adoption. Second, it identifies spatially based entrepreneurship ecosystems with high potential and analyzes some of the key challenges they face in generating more firms, including both DT-supplying and DT-adopting firms. Third, it discusses the key policies to address the main obstacles associated with technology adoption and start-up entrepreneurship and proposes how DT solutions could be used to facilitate this process.

TECHNOLOGY UPGRADING: TOWARD BETTER FIRMS

The findings presented in this section are based on the new Firm-Level Adoption of Technology (FAT) survey in Senegal, as well as on a survey of DT adoption by micro enterprises based on Research ICT Africa data.[10] The FAT survey—developed by Cirera, Comin, and Cruz (2020)—is a new tool to measure technology adoption at the firm level. The FAT data provide new measures of firm-level adoption along three dimensions: (a) standard measures of technologies; (b) technologies applied to general business support functions; and (c) sector-specific technologies. The standard, firm-level measures of technologies refer to "traditional" measures of general purpose technology (GPT) adoption, which enable firms to apply more technologies toward specific tasks. They include access and usage of GPTs such as electricity, phone, computers, internet, and social networks. Technologies applied to general business support functions are those technologies used by any firm, irrespective of the industries they are in, such as technologies used for business administration, production planning, sales, and payments methods. The sector-specific technologies are those applied to business functions that are industry specific, for example, land preparation in agricultural industries or input testing in the food processing industry. A detailed description of these technology measures—comparing data from Senegal, Vietnam, and the State of Ceará in Brazil—is provided by Cirera et al. (2020).

Standard measures of DTs: Use of information and communication technology

Except for basic mobile phones for business purposes, standard measures of DT adoption that focus on the use of information and communication technology (ICT) suggest that firms in Senegal are lagging. Although there is a clear correlation between the size of firms and access to landline phones, about 90 percent of firms use some type of mobile phone for business purposes (table 3.2). This finding is consistent with previous findings on the diffusion of fixed-line telephones compared with mobile phones in several countries in Africa. However, the same pattern is not observed with other digital enablers such as the number of computers, smartphones, and tablets. With regard to the quality of fixed internet access, most firms that use it rely on DSL (20 percent of all firms, or 63 percent of the 34 percent of firms that use the internet). About 12 percent of firms using fixed internet still rely on the much slower dial-up services. This number is larger for informal and small firms. The adoption of computers and smartphones or tablets for business purposes varies in expected ways across firm sizes. Larger firms have a significantly larger number of devices available, which is consistent with their scale. On average, small firms have less than 1 computer per firm (0.8), while there are about 2.7 computers per firm across all firms, and large firms have about 31 computers, either desktop or laptop, per firm.

TABLE 3.2 **Access and quality of ICT, by enterprise type**

TECHNOLOGY	MEAN (%)	STD. DEV. (%)	SMALL (%)	MEDIUM (%)	LARGE (%)	FORMAL (%)	INFORMAL (%)
Share of firms (%)							
Having telephone	32	47	26	55	76	87	17
Having mobile phone	89	32	89	88	90	88	89
Having computer	36	48	30	58	77	93	20
Having smartphone	30	46	28	36	40	36	28
Having internet	34	47	28	58	73	87	19
Type: Dial-Up internet	12	33	16	6	1	3	25
Type: DSL internet	63	48	55	75	87	85	34
Type: Wireless internet	12	33	13	13	8	10	15
Type: BPL internet	2	13	3	0	1	1	4
Acquisition of software	7.1	25.7	5.0	13.0	30.1	25.1	2.0
Average number of equipment							
Number of telephones	1.0	6.6	0.4	1.8	11.2	3.6	0.3
Number of mobile phones	3.9	13.1	2.2	7.1	31.9	9.3	2.4
Number of computers	2.7	24.4	0.8	5.5	33.5	10.7	0.3
Number of smartphones	0.9	11.2	0.5	1.6	8.4	2.6	0.4

Source: Cirera et al. (2021) based on the Firm-Level Adoption of Technology Survey data for Senegal.
Note: When internet type does not sum to 100, the difference is in the "others" category. Std. Dev. = standard deviation.

FIGURE 3.1

Access to the internet, own website, social media, and cloud computing, by size groups

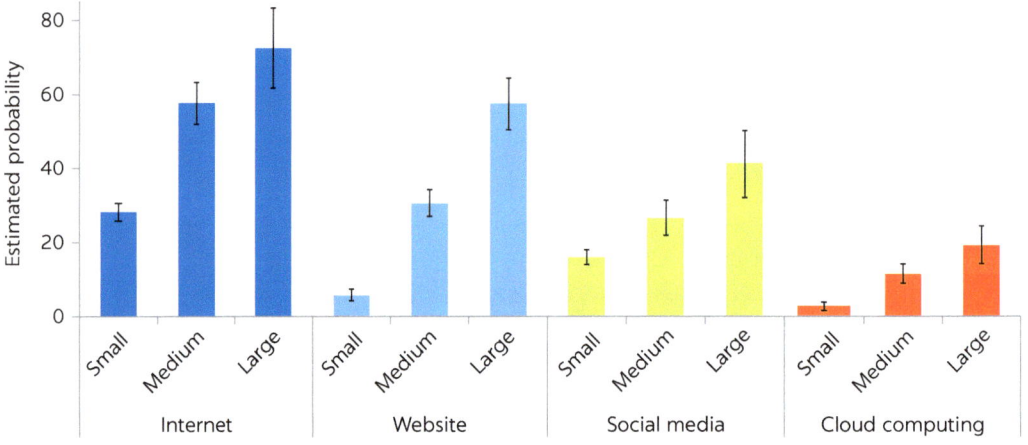

Source: Cirera et al. 2021.
Note: Estimated probability for using the respective technology, controlling for formality, sector, size, and regions.

Large firms are more likely to have access to the internet, a website, and a social media account, and use cloud computing, but the share of large firms adopting these DTs is still low. Fewer than 35 percent of establishments with 5 or more employees in Senegal have access to the internet, although it is more widely available among formal firms and large firms. Yet it is striking that about one-quarter of large firms do not have access to the internet. The share of large firms having their own website, using social media, and using cloud computing for business purposes is higher than for small and medium firms (figure 3.1).

These technology enablers are important, but these measures do not tell us how DTs are being used by firms. The main issue with these measures is that they do not identify the main purpose of use. For example, if the firm does have access to the internet or social media, how are these technologies being used to perform specific business functions? Overall, the lack of access to these GPTs becomes a constraint for the use of any DT solution that rides on top of the internet (for example, if a firm does not have a computer, it will likely not have access to an ERP system for production planning).[11] But having access to digital GPTs is not a sufficient condition for applying these technologies to perform specific tasks. The next subsection provides further details on how firms are using DTs applied to specific business functions.

Technology use for general business functions

Advanced technologies for performing general business functions (GBFs) are predominantly digital. GBF tasks are defined as those being commonly performed across all firms, irrespective of the industries they are in. Figure 3.2 presents the distinct technologies associated with six GBFs that the survey covered, which are (a) business administration, (b) production planning, (c) obtaining customer information for marketing and new product development, (d) sales methods, (e) payment methods, and (f) quality control. The technologies associated with each of these GBFs follow a ladder of sophistication that goes from the most basic (for example, a handwritten process) to the most sophisticated level (for example, ERP systems for business administration).[12]

Most firms in Senegal still rely on predigital technologies to perform GBFs. Even when firms are adopting DTs to perform GBF tasks, they are still very basic. A very small number of firms, fewer than 10 percent, use specialized software or ERP systems for processing information related to business administration (accounting, finance, and human resources). The use of computers with standard software is the most diffused DT for business administration, but still fewer than half of the 29 percent of firms that perform this task extensively with standard software rely intensively on it (13 percent).[13]

Among GBFs, payment method is an exception, with a large share of firms already adopting more advanced DTs, but only on the extensive margin. The results summarized in figure 3.2 indicate that a large majority of firms still rely mainly on the most basic technologies to perform these tasks: 77 percent rely mostly on handwritten methods of business administration; 91 percent rely mostly on face-to-face chats to obtain consumers' information for marketing purposes and product development; 96 percent rely mostly on the establishment's premises or phone, email, and representatives for sales; and 85 percent rely mostly on cash for payments.

For measurement of the technological gap, the technologies are combined into an index for both extensive and intensive margins, summarizing the technological sophistication for each business function.[14] The index varies between 1 and 5, where 1 stands for the most basic level of technology and 5 reflects the most sophisticated. With the help of experts for each industry, a rank was assigned to the technologies in each business function according to their sophistication and the complementarity or substitutability within business functions.[15] Figure 3.3 shows the average firm-level technology index for each GBF. The results emphasize important differences between extensive and intensive margins and across GBFs. Although some firms are adopting more sophisticated

FIGURE 3.2

Share of firms using technologies applied to GBFs, extensive and intensive margins

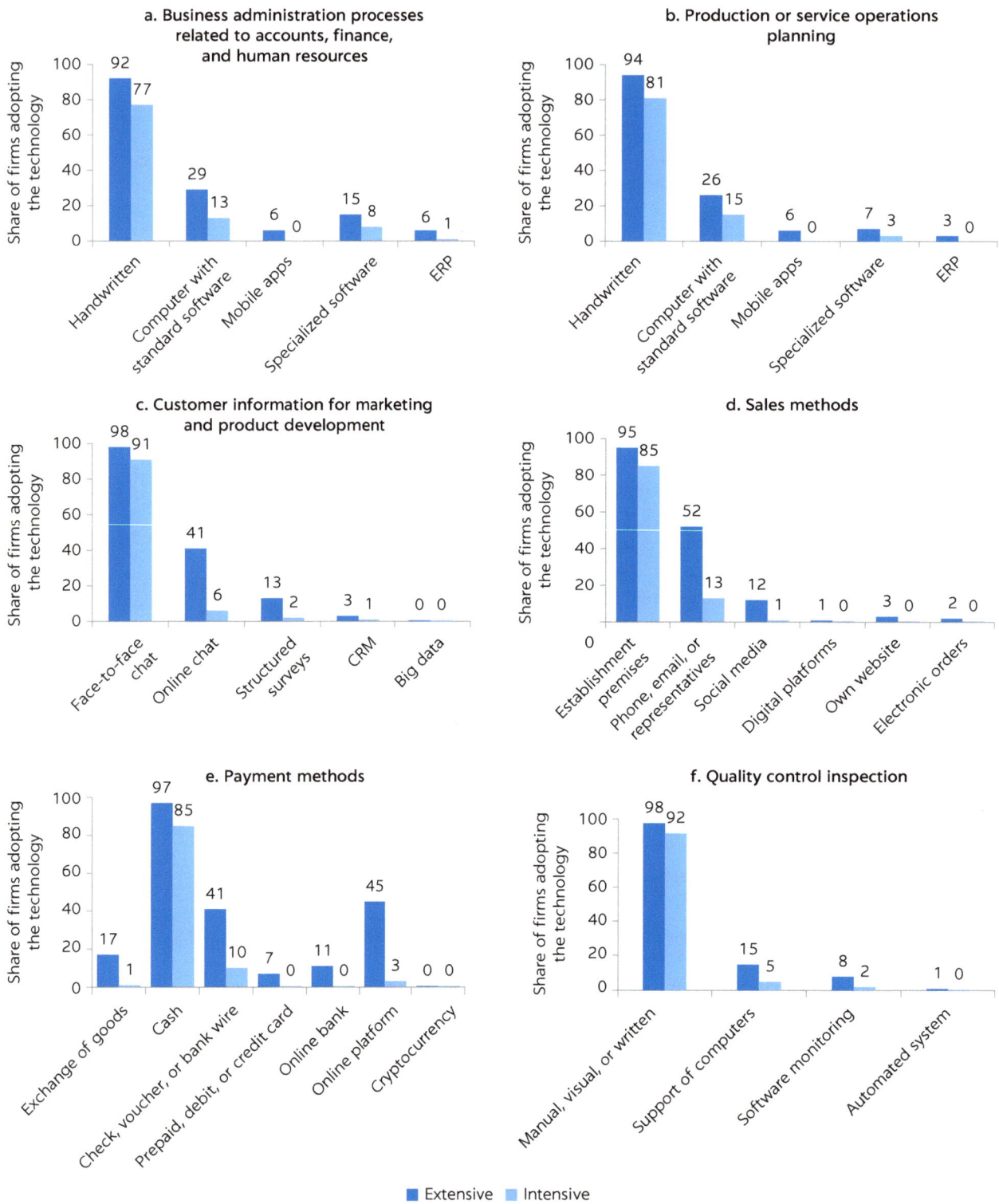

Source: Cirera et al. 2021.
Note: The extensive margin refers to the share of firms using a given technology for a given business function (whether they use it at all). The intensive margin refers to the share of firms using this technology as the most frequently used technology to perform this business function (whether they use it the most). CRM = Customer Relationship Management; ERP = Enterprise Resource Planning.

technologies for a specific GBF (for example, payment methods), these are not the most-used technologies, and the gaps are large, especially in formal firms. The gap between the extensive and intensive margins also varies across GBFs.

Formal firms in Senegal are more advanced than informal firms in technology adoption, but they still face a large gap. This gap is statistically significant and robust across different GBF dimensions, even after controlling for size, sector, and region. Figure 3.4 shows that formal firms dominate informal firms on the adoption of technologies in all GBFs. But the gap decreases significantly when considering the intensive margin. Although, for payment methods, both formal and informal firms use relatively advanced technologies (for example, digital payment methods), both groups rely mostly on simpler technologies for other GBFs. In business administration (for example, human resources and finance), formal firms are more likely to use them in the extensive margin, but also have a large gap in the intensive margin. Large firms also use more advanced technologies in most GBFs, but the gap across size is smaller for the intensive margin. This is because firms in general adopt less-sophisticated technologies.

Firms in services and manufacturing are adopting more advanced technologies for GBFs, but their levels of adoption are also low and the differences in the intensive margins are small. Similar patterns of more advanced technologies for payment methods (for example, e-wallet and digital platforms) exist in the extensive margin, but overall large gaps in the intensive margins are also observed across sectors. Agriculture has a larger share of informal firms and lower levels

FIGURE 3.3
The technology index applied to GBFs

Source: Cirera et al. 2021.

FIGURE 3.4
Levels of technological sophistication, by GBF: Formal vs. Informal firms

Source: Cirera et al. 2021.

of technologies applied to GBFs than other sectors. No matter the sector of the firms, only a very small share has adopted frontier technologies. Although the fourth industrial revolution is a common term in many policy discussions,[16] firms in Senegal have yet to widely adopt the technologies associated with the second and third revolutions. Almost 80 percent of firms with more than 5 employees in manufacturing still rely mostly on manual procedures for fabrication (intensive margin). Even in the extensive margin, only a small share of firms (fewer than 10 percent) is adopting machines operated by computers; the use of robots or 3D printers is not observed in the intensive margin in Senegal.

Technology use for sector-specific business functions

Sector-specific business functions (SSBFs) are used to determine the level of sophistication of technologies related to core production processes or services by specific industries. Overall, agricultural firms that produce crops, vegetables, and fruits in Senegal use a low level of technological sophistication, both on extensive and intensive margins (table 3.3). The average SSBF index for Senegal is 1.6, while the intensive margin is 1.3. When the index is split across four sectors for which the survey is stratified—agriculture, food processing, wearing apparel, and wholesale/retail—the extensive margin varies between 1.58 in wholesale/ retail and 2.03 in food processing, while the intensive margin varies between 1.21 in wholesale/retail and 1.48 in food processing.

In agriculture (crops, fruits, and vegetables), the SSBFs with more advanced technologies are those related to land preparation and irrigation, both on intensive and extensive margins. These results suggest that farms on average are using technologies that go beyond manual operation for these tasks, such as animal-aided tools or tractors for land preparation (figure 3.5, panel a). Similarly, for irrigation, technologies such as surface flood irrigation are adopted by 13 percent of the firms, and irrigation by small pump is adopted by 21 percent on the extensive margin; about 20 percent of firms are also adopting one of these two technologies on the intensive margins. The small gap between extensive and intensive margins in irrigation suggests that this SSBF is one in which a business function has been adopted intensively and is the most advanced technology used by the farmer. Yet for tasks related to packaging, storage, harvesting, weeding, and pest management, the level of technological sophistication is below 2 on the extensive margin, and below 1.5 on the intensive margin. Overall, these results suggest that farms in Senegal are still relying mostly on manual operations to perform these tasks. This fact suggests heterogeneity in the level of technology used across SSBFs within farms.

In food processing, the technology sophistication index for SSBFs varies between 2.6 for mixing, blending, and cooking or food storage and 1.8 for input testing on the extensive margin. These values are smaller on the intensive margin, varying from 2.1 for mixing, blending, and cooking to 1.3 for input testing

TABLE 3.3 **Levels of technological sophistication across SSBFs, by sector**

SSBF TYPE	AGRICULTURE	FOOD PROCESSING	WEARING APPAREL	WHOLESALE AND RETAIL
Extensive	1.75	2.03	1.58	1.58
Intensive	1.30	1.48	1.34	1.21

Source: Cirera et al. 2021.
Note: SSBF = sector-specific business function.

(figure 3.5, panel b). Overall, the indexes on the extensive margin suggest that the average firm has been adopting technologies with the support of machines, although many machines are still manually operated. Although almost 30 percent of establishments rely on supplier reviews for input testing,[17] 86 percent of establishments rely on "human sensory" methods, which is the most basic procedure available to perform this task. For mixing, blending, and cooking, firms are using machines, but mostly manually operated ones. Seventy-three percent of firms use "manually operated machines" on the extensive margin, and about half of firms use this technology in the intensive margin, while 32 percent of firms still rely on fully manual processes. For antibacterial processes, the average firm has adopted between "wash or soaking" and "thermal methods," but 43 percent still rely mostly on minimal processing or wash or soaking. For packaging, most firms (85 percent) still use manual procedures as the more-often-used technologies; while for storage, 82 percent rely on the most basic technologies (minimal protection or a closed building).

For wearing apparel, most firms still rely on manual design and cutting, manually operated machines for joining parts, and manual ironing as the most-used technologies. Sewing is the business function with the highest index, for which a large share of firms is adopting manually operated machines (almost 80 percent) or semiautomatic sewing machines (22 percent) (figure 3.6). Sixty-eight percent of firms use one of these processes as the most-used technology for sewing. On the other hand, design and cutting are the business functions for which most firms still rely on manual processes. Ninety percent of firms use manual cutting in the intensive margin; almost all firms rely only on manual design without adopting digital technologies such as digital 2D or CAD.

In retail, on average, firms are still relying mostly on manual technologies for customer services, pricing, merchandising, inventory, and advertisement. About 20 percent of firms are using social media for customer service, but less than 3 percent of them use it as the most frequently used technology (figure 3.6).

FIGURE 3.5

Levels of technological sophistication, by SSBF in agriculture and food processing

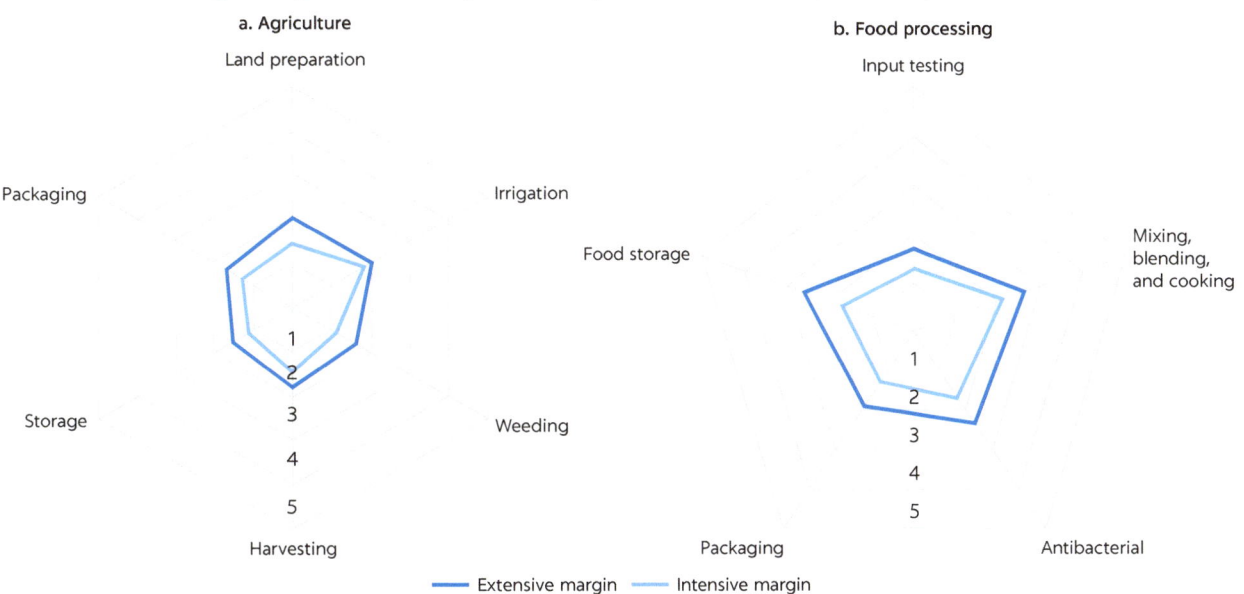

Source: Cirera et al. 2021.

FIGURE 3.6

Levels of technological sophistication, by SSBF in apparel and retail

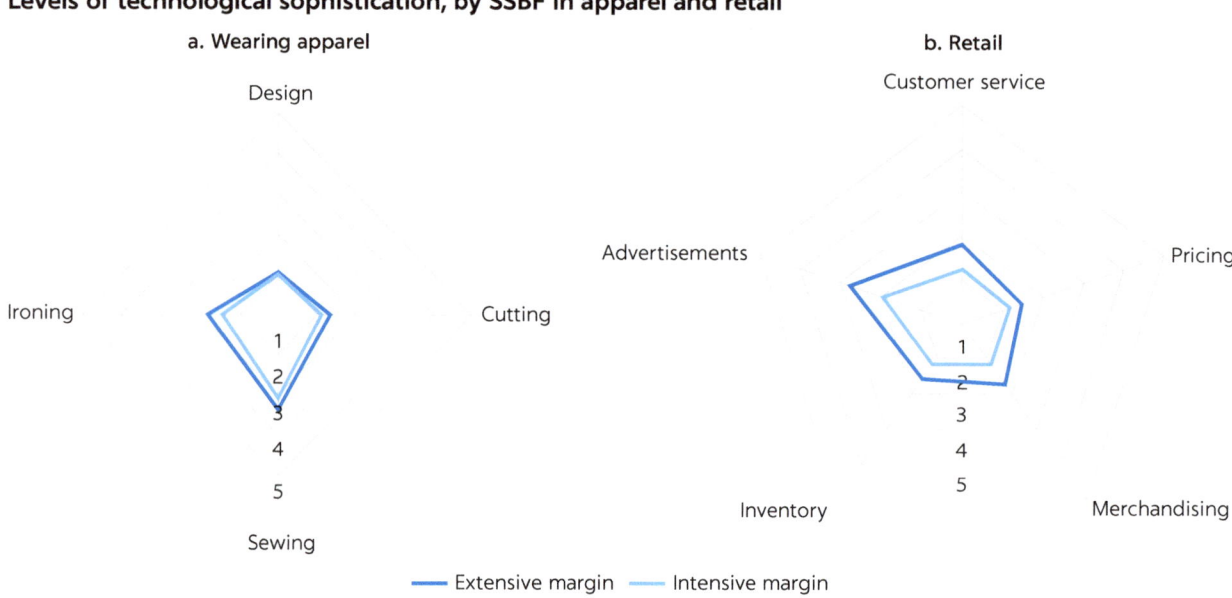

Source: Cirera et al. 2021.

In the intensive margin, almost 95 percent of firms provide the services at the premise (83 percent) and/or by phone (13 percent). Eighty-two percent of firms are relying mostly on handwritten records for inventory; and 63 percent rely mostly on paper-based communication or radio, billboards, and TV as the most frequent technologies for advertisements. DTs are relatively more relevant in the extensive margin for advertisements, where 43 percent of firms also use email or mobile phones and social media.

Technological gaps and productivity

Compared with the state of Ceará in Brazil, firms in Senegal have a gap of 36 percent and 30 percent on the extensive and intensive margins, respectively, for performing general business functions.[18] This fact means that the average firm in Senegal is significantly behind the average firm in one of the poorest states in Brazil in terms of the sophistication of technologies used to perform basic business functions, like accounting and accepting payments for the sale of a good or service. The average Senegalese firm is likely to keep handwritten records for accounting and accept cash for payments, while the average firm in Ceará is more likely to use digital solutions among the technologies it uses, and also for the technology it uses most often. A large relative gap, though somewhat smaller, is also observed when comparing the average firm in Senegal with the average firm in Vietnam (table 3.4).

Despite the differences in the average across countries, there is significant heterogeneity in technology adoption across firms and across business functions in Senegal. Figure 3.7 shows that informal, small, and agricultural enterprises lag their respective groupings. Importantly, even formal and large firms lag the average levels of technological sophistication across all firms of global comparators: the unconditional average for the intensive margin of large Senegalese firms is at 1.85, relative to the average across all firms in Vietnam of 1.9 and 2.5 for Brazil.

TABLE 3.4 **Technological gaps between Senegal and other countries**

COUNTRY	GENERAL BUSINESS FUNCTIONS		SECTOR-SPECIFIC BUSINESS FUNCTIONS	
	EXTENSIVE	INTENSIVE	EXTENSIVE	INTENSIVE
Ceará (Brazil)	3.4	2.5	2.8	1.9
Vietnam	2.8	1.9	2.6	1.8
Senegal	1.9	1.3	1.6	1.3
Gap: BR – SN	1.5	1.2	1.2	0.6
Relative gap[a]	36%	30%	29%	16%

Source: Cirera et al. 2021.
a. Relative gap is the difference between Brazil and Senegal relative to the maximum technology gap of 4: (Ceará-Brazil – Senegal) / 4 (maximum gap). Additional decimal points were used to calculate the relative gap.

FIGURE 3.7

Levels of technological sophistication across GBFs, by firm types (conditional)

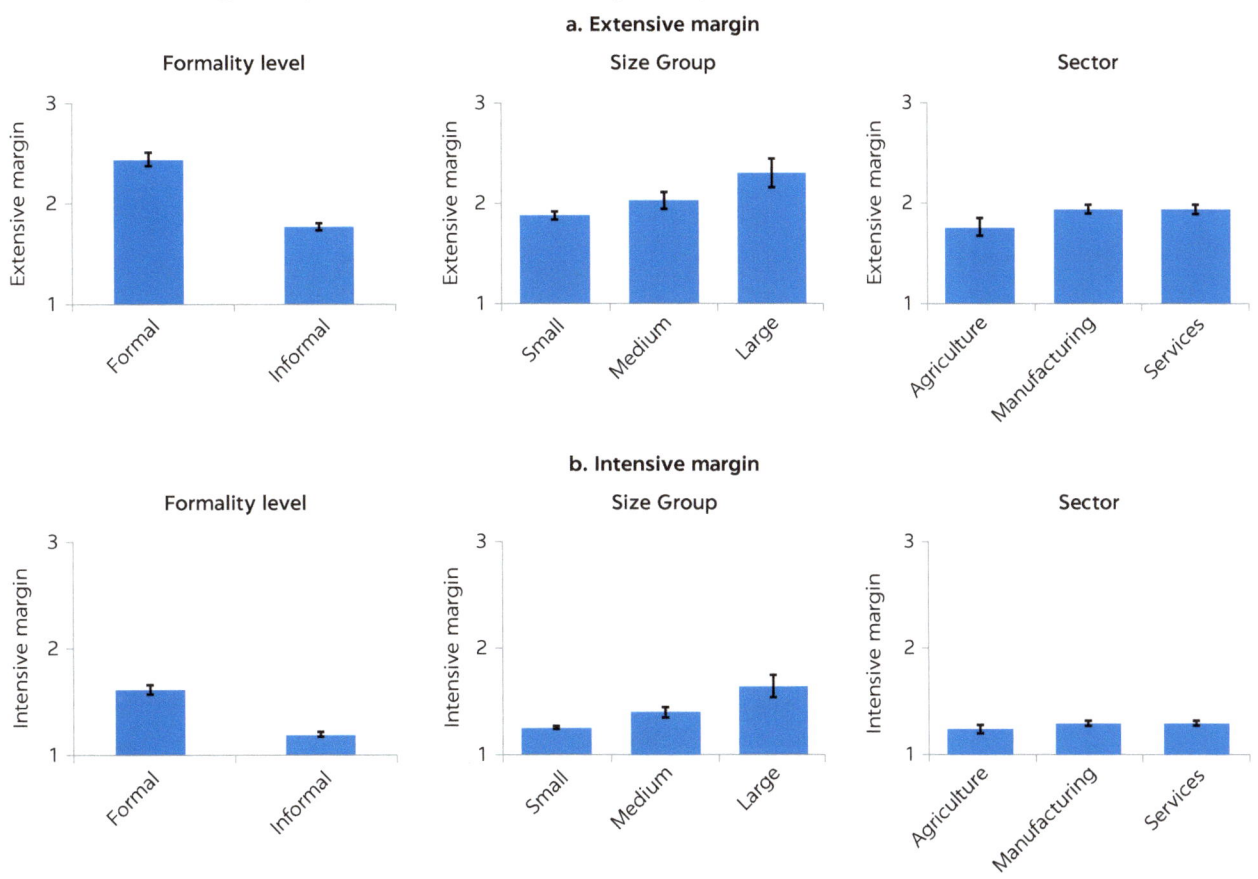

Source: Cirera et al. 2021.
Note: Results are based on predicted values from estimates conditional on formal status, size, and sector groups.

Therefore, despite a significant heterogeneity across firms within the country, the relatively better firms of Senegal (formal and large) still have a lot to catch up on in terms of technology adoption. For example, a very small share of firms in Senegal is benefiting from frontier technologies, such as those associated with Industry 4.0 (see box 3.1).[19]

How far are manufacturing firms in Senegal from Industry 4.0?

Most Senegalese firms are not adopting Industry 4.0 technologies yet. Only a very small share of firms in Senegal has been using so-called Industry 4.0 technologies (figure B3.1.1). Among those technologies is cloud computing, which is the most diffused Industry 4.0 technology in Senegal and is used by fewer than 5 percent of firms. More autonomous technologies, such as AI, robots, 3D printers for manufacturing, and precision agriculture, are used by fewer than 1 percent of Senegalese firms.

Adoption of Industry 4.0 technologies

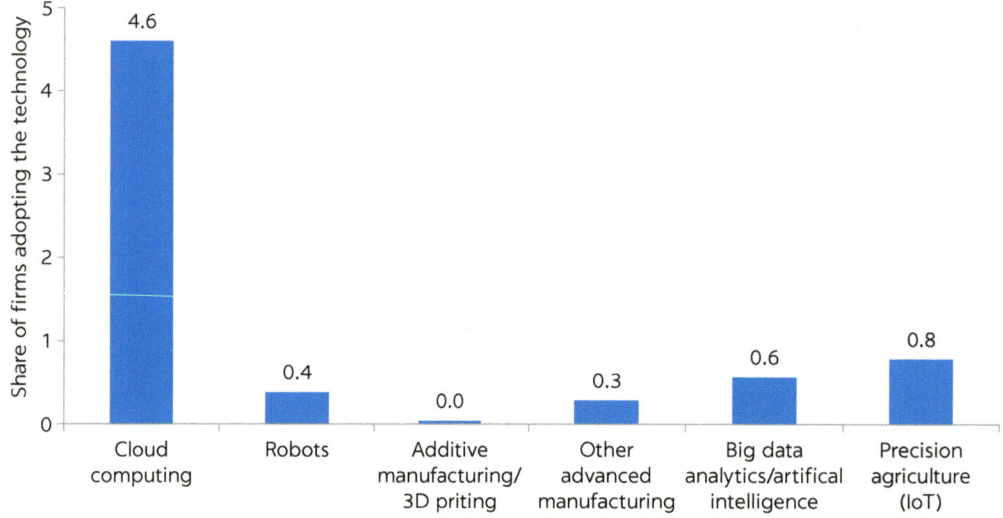

Source: Cirera et al. 2021.
Note: IoT = internet of things.

Differences in technology adoption across countries, regions, sectors, and firms are positively associated with productivity. New evidence from the FAT survey across countries identifies a statistically significant correlation—from 0.83 to 0.92—between different measures of adoption (for example, GBFs and SSBFs, extensive and intensive margins) and regional productivity at the subnational level for Senegal, Vietnam, and the Brazilian state of Ceará.[20] Moreover, they show through development accounting exercises that variations in technology adoption across firms account for about one-third of the observed differences in productivity across firms and about 25 percent of the agricultural-versus-nonagricultural gap in cross-country differences in productivity, based on comparisons of Senegal, Vietnam, and the state of Ceará.

The strong association between technology adoption and productivity, for both extensive and intensive margins, is also observed for Senegalese firms. An analysis of the relationship between technology adoption and firm performance (labor productivity or value added per worker) reveals a positively and statistically significant correlation in a sample of firms in Senegal (figure 3.8). Although these results do not suggest any causal relationship between technology and

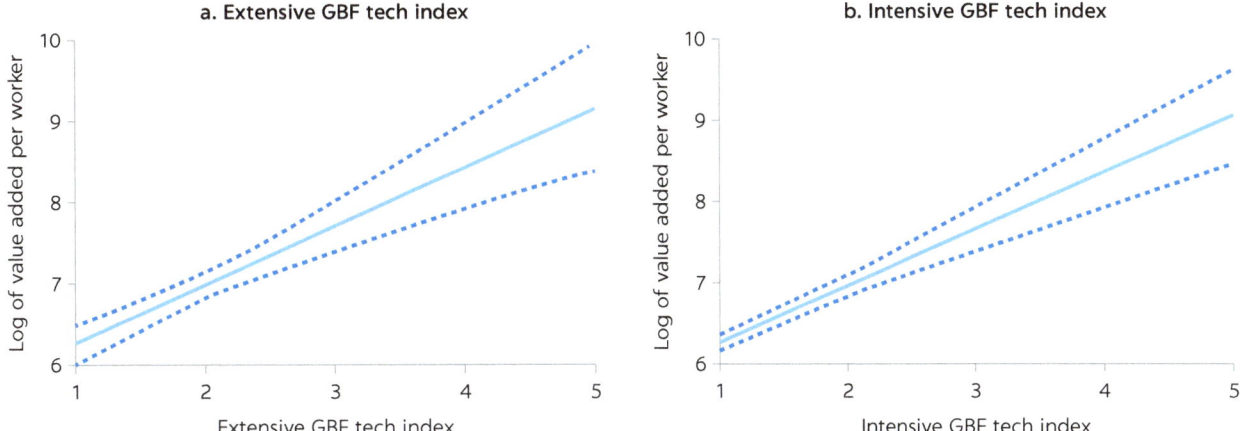

FIGURE 3.8

Firm-level tech adoption index and value added per worker

Source: Cirera et al. 2021.
Note: These results are based on linear estimations of the relationship between log of the value added per worker and the GBF technology index, controlling for size, sector, region, and informal status.

performance, the existing literature (Atkin et al. 2017; Comin and Hobijn 2010; Comin and Mestieri 2018) has shown that technology is an important driver of productivity, based either on cross-country analyses using structural models or on randomized control trial experiments.

Main barriers to generating better firms

Among Senegalese firms, a large share of firms reports that financial constraints and lack of information or knowledge are key perceived obstacles for adopting better technologies. The FAT survey asks firms about their top three obstacles to adopting technology. Figure 3.9 provides the share of firms reporting obstacles by firm size group. Financial constraints and lack of firm capabilities—including access to information, knowledge, and technical capacity—are among the top perceived obstacles for firms (reported by approximately 70 percent of firms), disregarding their size group, although the order of importance changes between these two obstacles. Among small firms, a larger share indicates financial constraints, while among medium and large firms, a larger share indicates the importance of lack of capabilities.[21]

Financial constraints, lack of firm capabilities, and uncertainty or lack of demand are among the main perceived obstacles to adopting better technologies.[22] First, the results indicate that technology adoption for GBFs or SSBFs, as well as on the extensive or intensive margins, are strongly and positively associated with firm size and negatively associated with informality status. The results also suggest that technology adoption for GBFs is negatively associated with lack of information, lack of knowledge, uncertainty, consumers' preferences, and financial constraints. For SSBF technologies, the perceived obstacles are associated with lack of information, consumers' preferences, and government regulations.[23]

The importance of these subjective obstacles is reinforced by objective data. Beyond its question on perceived obstacles, the FAT survey also provides

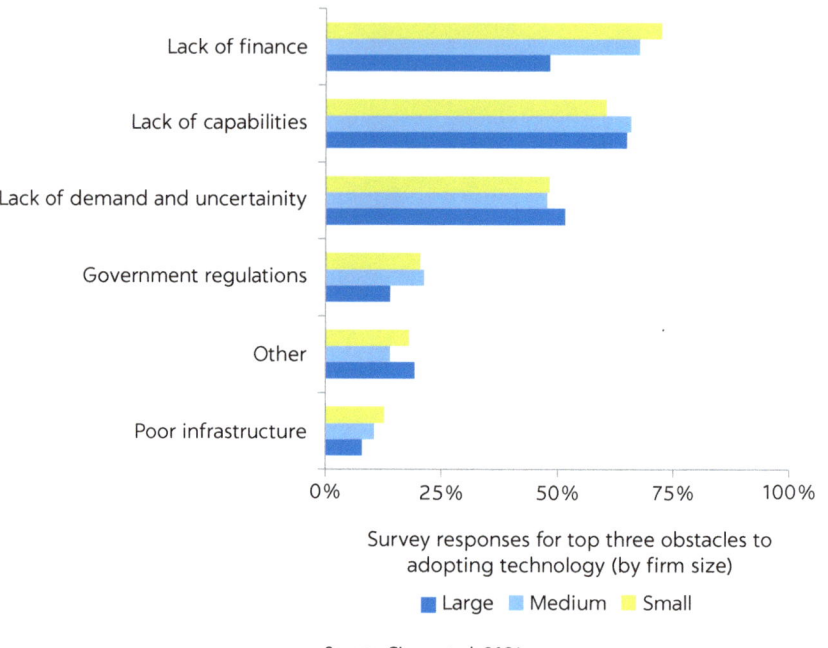

FIGURE 3.9

Perceived obstacles for adopting technologies

Survey responses for top three obstacles to
adopting technology (by firm size)

■ Large ■ Medium ■ Small

Source: Cirera et al. 2021.

information that can be used as a proxy to measure issues associated with financial constraints, information and knowledge, and challenges related to access to markets. Overall, the results support small and informal firms being more likely to face these challenges. Factors such as having business interactions with larger or multinational firms or having top managers with a college degree and education abroad as an important source of information and knowledge are significantly associated with adoption.[24]

In line with these results, informal and smaller firms face higher interest rates and have less access to loans for purchasing machines or software. The likelihood of firms taking loans to purchase machines or software is positively associated with higher levels of technology adoption (figure 3.10). The FAT survey also suggests that the average interest rate is lower for formal and large firms. Similarly, the share of firms taking loans increases from small firms (18 percent) to large firms (40 percent), while the average interest rate decreases. These results suggest that informal and small firms experience further challenges to access finance, which can constrain them from adopting better technologies (Abate et al. 2016; Bircan and De Haas 2020; Cole, Greenwood, and Sanchez 2016; Midrigan and Xu 2014).

Another important source of knowledge is associated with the level of human capital of managers and workers. Figure 3.11 presents bar charts of advanced degrees of top managers and workers by formality and firm size. Although 73 percent of top managers in formal firms have a bachelor's degree or higher level of education, less than 30 percent of top managers in informal firms have a bachelor's degree or more. There is also a significant positive relationship between size and top manager skills. A similar relationship is observed for workers with vocational education, engineering or applied science,

FIGURE 3.10

Technology adoption and access to finance

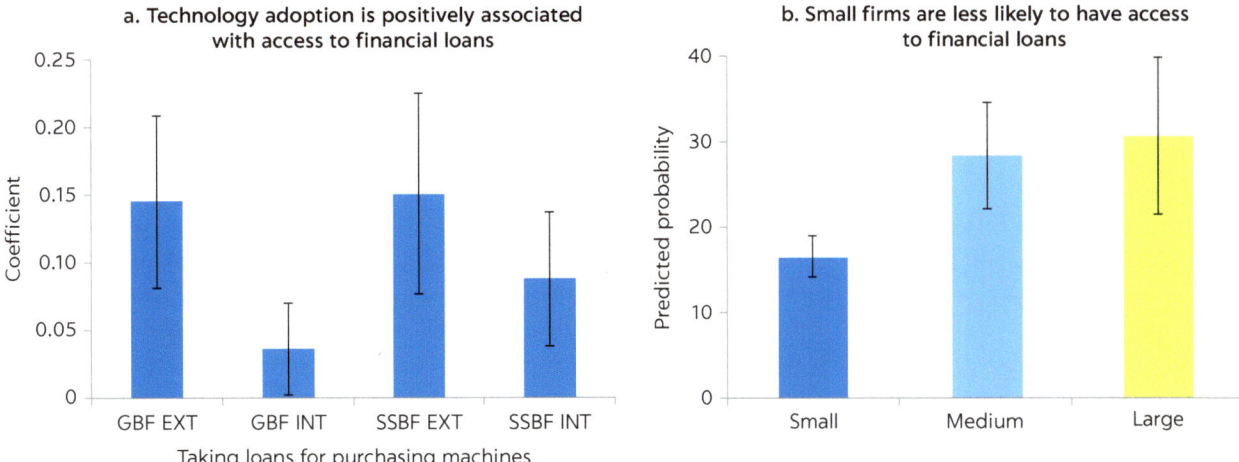

a. Technology adoption is positively associated with access to financial loans

b. Small firms are less likely to have access to financial loans

Source: Cirera et al. 2021.
Note: Point estimates at 5 percent confidence intervals. Panel a shows the coefficient for each technology measure regressed on a dummy for taking loans to purchase machines or software, while controlling for formality, sector, size, and regions. Panel b shows the predicted probability of getting loans by size groups and confidence intervals from the probit regression with controlling for other baseline characteristics.

FIGURE 3.11

Technology adoption and firm capabilities

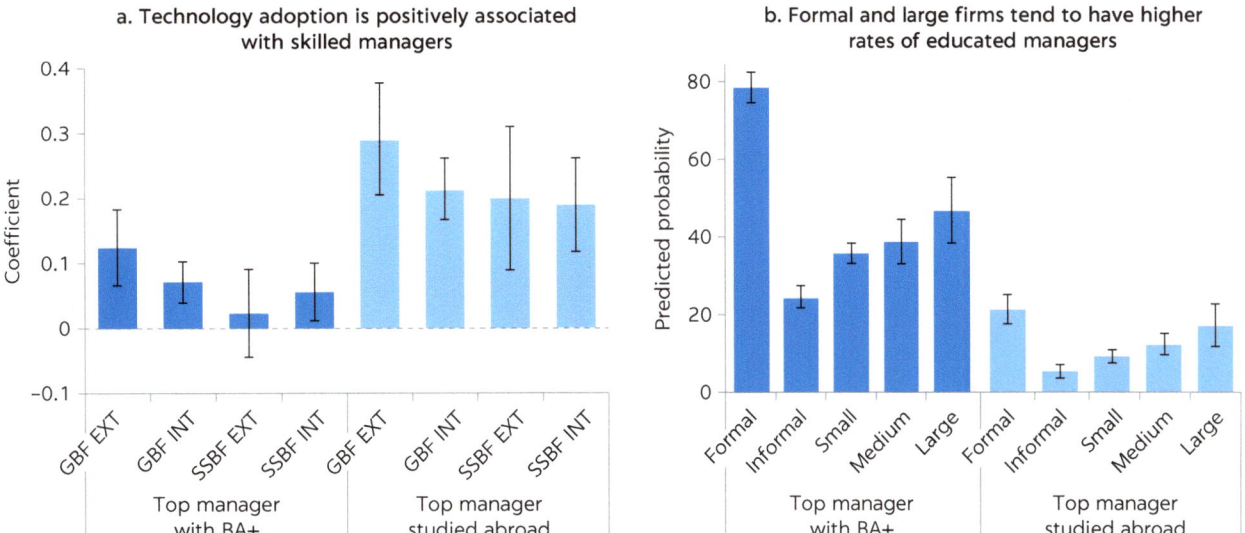

a. Technology adoption is positively associated with skilled managers

b. Formal and large firms tend to have higher rates of educated managers

Source: Cirera et al. 2021.
Note: Point estimates at 5 percent confidence intervals. Panel a shows the coefficient for each technology measure regressed on a dummy for top managers' education (for example, BA+ and studied abroad) and the percentage of workers with different education levels (for example, secondary school, vocational training, and college degree), respectively, while controlling for formality, sector, size, and regions. Panel b shows the predicted probability of having top managers with BA+ or studying abroad by formality and size with confidence intervals from the probit regressions, controlling for other baseline characteristics.

master's, MBA, and PhD degrees. Formal and large firms have a much larger share of educated workers than informal and small firms (Caselli and Coleman 2001; Comin and Hobijn 2010).

Participation in global value chains (GVCs) is also associated with adoption of more advanced technologies. Firms with higher levels of technologies are significantly more likely to trade in both ways, as an importer and exporter (figure 3.12). Although export-import status is more likely among large firms, the association between technology and two-way trade is robust for all combinations of technology measures (GBFs, SSBFs, extensive, and intensive) controlling for formality status, sector, region, and size. A higher level of adoption is associated not only with the interaction through international trade but also with the business interaction with multinational enterprises (MNEs), as a supplier or a buyer. As suggested by previous literature, firms can benefit from information flows and exposure to better business practices when exposed to the competition and opportunities associated with trade.

In addition to the strong association with two-way trade, digital technologies can also be used to facilitate trade procedures. Senegalese firms face high uncertainty in delays to import goods (see box 3.2). Senegalese manufacturing firms indicate that they experience a relatively high number of days to clear goods at the border and face high uncertainty in such delays, as reported in World Bank Enterprise Surveys. These figures are confirmed by customs transaction data for Senegalese importing firms for the period 2015–18. Yet uncertainty over the time required to clear the consignment after it has reached its destination can lead to postponed delivery of key inputs, thereby disrupting the supply chain and increasing the risk for contract termination from the international buyer.

FIGURE 3.12

Technology adoption and trade

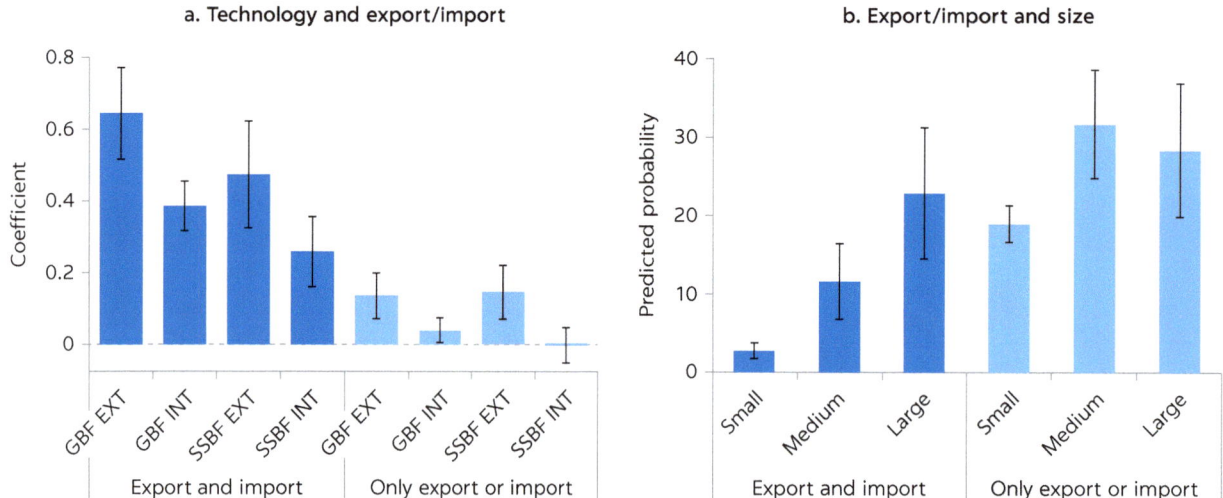

Source: Cirera et al. 2021.
Note: Point estimates at 5 percent confidence intervals. Panel a shows the coefficients for each technology measure regressed on exporter/importer dummies, respectively, while controlling for formality, sector, size, and regions. Panel b shows the predicted probability of exporter/importer status on size from the probit regressions, while controlling for other baseline characteristics.

BOX 3.2

How DTs could further support trade facilitation in Senegal

Evidence reveals that the import time uncertainty that Senegalese firms in global value chains (GVCs) face harms their export survival, as well as their export value and quality. Results based on customs transaction data for export-import firms in 2015–18 in Senegal indicate that uncertainty in how long it will take for imported goods to be available imperils the survival of firms' new export flows, by reducing the chance that a firm can continue to serve a new market or export new products the following year (figure B3.2.1). Furthermore, if the uncertainty in the time required to complete customs procedures and border formalities could be reduced by 10 percent (in time), then the exported value could increase by 4 percent and the unit value by 1 percent (as a proxy for export quality or sophistication). These results for Senegal are in line with findings for 48 developing countries (Vijil, Wagner, and Woldemichael 2019).

Digital tools, such as tracking and tracing mechanisms or risk management models, could promote export performance and GVC participation by enhancing predictability. The findings support stepping up investments designed to reduce supply chain unreliability caused by unpredictable import clearance times. Investments would include incentives for border control agencies, port operators, and other transportation and logistics stakeholders to adopt information technology and electronically interconnect themselves to address coordination failures among public and private actors involved in the movement of goods. Effectively implementing the World Trade Organization (WTO) Trade Facilitation Agreement, especially provisions on advance rulings and border agency and customs cooperation, should also increase predictability and reduce border clearance times. Expanding the Authorized Economic Operator Program would also open a predictable clearance channel for compliant firms, which are likely to be those participating in GVCs.

Uncertainty between vessel arrival at the port and the registration of the customs declaration seems to be triggering this economic harm of import time uncertainty for exporters. Delays attributable to port operators' activities instead of delays at customs or a lack of cash flow from the importer to pay import duties seem to drive this effect. Furthermore, results do not seem driven by the behavior of specific firms changing over time (controlling for product-specific import time instead of firm-specific time) nor their brokers (controlling for brokers). Overall, results

FIGURE B3.2.1

Import delays between vessel arrival and registry of customs declaration

Source: Vijil 2020.

continued

Box 3.2, *continued*

suggest that port service inefficiencies are among the key drivers of high import time uncertainty in Senegal. This finding is in line with Senegal's position in the Logistics Performance Index. Results suggest that these port inefficiencies also affect firm export performance and participation in time-sensitive GVCs.

Senegal pioneered setting up an electronic Trade Single Window in 2004 that 12 years later had 71 public and private connected partners. An import transaction involves multiple public and private stakeholders with different objectives and incentives, as well as multiple exchanges of documents. Each interaction increases the risk of delays and becomes a source of uncertainty. The Trade Single Window system allows a single submission of data for multiple use, instantaneous data transfer to multiple actors, gains in security, and reduced process time. It is estimated that the

time required to complete preclearance formalities has been reduced by 70 percent, from 4 days to only half a day, since 2004. The time associated with the clearance process has been cut by 50 percent, from an average of 18 days to just 9 days (UNECE 2016).

Senegal could still leverage DTs to further advance the trade facilitation agenda on border agency cooperation, customs processes, and automation. Senegal performs relatively better than average among regional and income group peers. However, it still lags Kenya in best practices on some of the WTO Trade Facilitation Agreement policy areas, especially border agency cooperation and customs processes (figure B3.2.2). On border cooperation of domestic agencies involved in cross-border trade management, the interconnection or sharing of computer systems as well as exchanging data in real time could significantly reduce average

FIGURE B3.2.2

Trade facilitation practices: Senegal and regional and income group peers

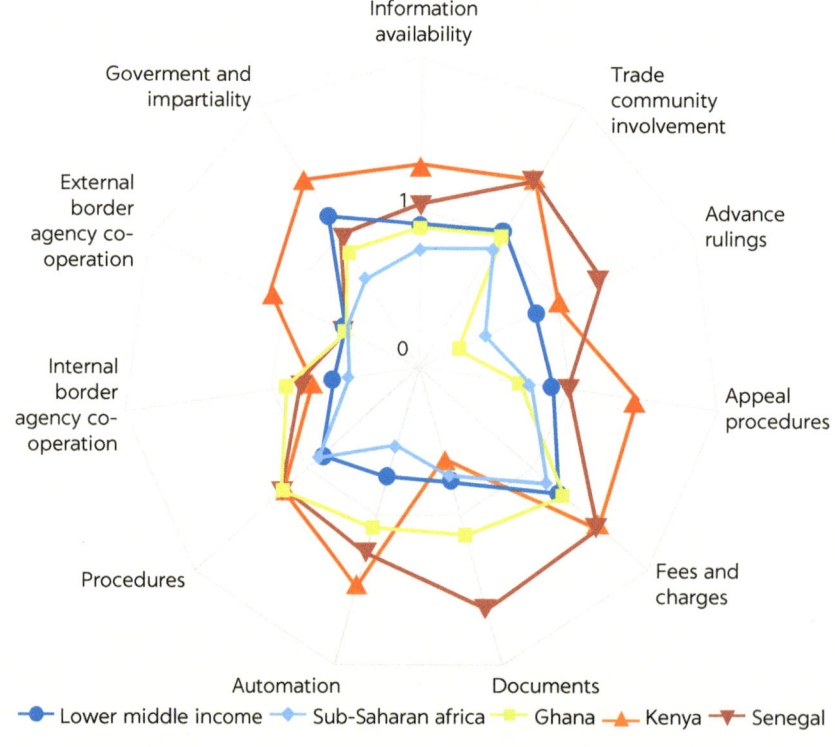

Source: Organisation for Economic Co-operation and Development (OECD) Trade Facilitation Indicators.

continued

Box 3.2, *continued*

times and uncertainty (and would be particularly beneficial for integrating into agribusiness GVCs). Regarding potential improvements on automation, the possibility to lodge documents in advance in electronic format for prearrival processing and allowing for the release of goods subject to conditions (that is, guarantee) also are priorities. In terms of customs procedures, perishable goods could benefit from preferential treatment by separating their release from clearance. Likewise, establishing standard policies and procedures to guide audits would reduce uncertainty for known traders in their access to goods.

The largest gains from DTs in competitiveness and participation in GVCs could arise from improving logistics performance; the operationalization of the Port Single Window and of the Port of Dakar Vehicle Booking System are of paramount importance. The Port of Dakar is the main commercial port of entry in Senegal. It plays a critical role in the competitiveness of the economy, both because it depends heavily on imports and because many of the most competitive companies rely on its operation for their exports. Senegal is still lagging peers in almost all dimensions of the 2018 World Bank Logistics Performance Index, especially in the ability to track and trace consignments and the timeliness of shipments in reaching destinations within the expected delivery time. Uncertainty in the time to import between vessel arrival and registration of the customs declaration, mostly related to logistics issues, is reducing firm export survival, value, and quality (Vijil 2020).

Finally, digitally enabled trade facilitation measures can contribute to Senegal's response to the COVID-19 crisis by expediting the movement, release, and clearance of goods, including those in transit. Access to medical and personal protection equipment, hygiene goods, and food has become highly uncertain because of production stops and other blockages along GVCs. Bottlenecks at border posts can put more pressure on the capacity of the health system and the population to respond effectively. While firms importing COVID-19 goods faced relatively lower import clearance times for these goods relative to other goods in 2018, firm-specific median import times and their uncertainty remain high. For instance, firms importing material for testing and case management (for example, enzymes, ventilators, and other medical equipment) faced relatively high median import times and significant uncertainty in the delays it took for goods to be cleared by border agencies. Strengthening the risk management system to allow low-risk critical supplies to go through expedited clearance controls should significantly reduce import delays. Because many of the COVID-19 products require import clearance from border agencies other than customs, increased internal and external border agency collaboration is key. Senegal also has a critical role to play in the management of the crisis through the region, considering that it is one of the main transit entry gates for Mali.

Sources: OECD; World Bank 2020; Vijil 2020.

Technology and jobs

Most firms report that they do not change the number of workers in response to digital adoption. When asked about how firms adjust their labor to the adoption of new technologies through the acquisition of new machines, equipment, or software, about 78 percent of firms suggest that they do not change the number of workers (figure 3.13), with 60.8 percent reporting that they do not implement changes and 27.3 percent suggesting that they offer some training to current workers. Only a small number of firms, about 2 percent, report a reduction in the number of workers as an adjustment for the acquisition of new technologies, which is a much smaller share than the number of firms that report an increase in the number of workers with the same skills (3.9 percent) or the hiring of more qualified workers (6.1 percent).

Firms with better technology also create more jobs. Figure 3.14, panel a, shows a positive and statistically significant association between employment growth (between 2016 and 2018) and technology adoption across different measures of technology, controlling for the initial size of the firm in 2016. The result is robust for different measures, including controlling for age, sector, region, foreign ownership, and exporting status of firms (figure 3.14, panel b). An increase of 1 point in the technology adoption index for GBFs that the firm uses most intensively, such as using standard Excel software rather than writing by hand for accounting and inventory control, is associated with a 14 percent increase in the number of workers in the average firm. A similar increase in technological sophistication at the extensive margin is associated with a 7-percentage-point increase in the number of workers in the average firm. The coefficient is smaller

FIGURE 3.13

How firms adjust jobs with the adoption of new technologies

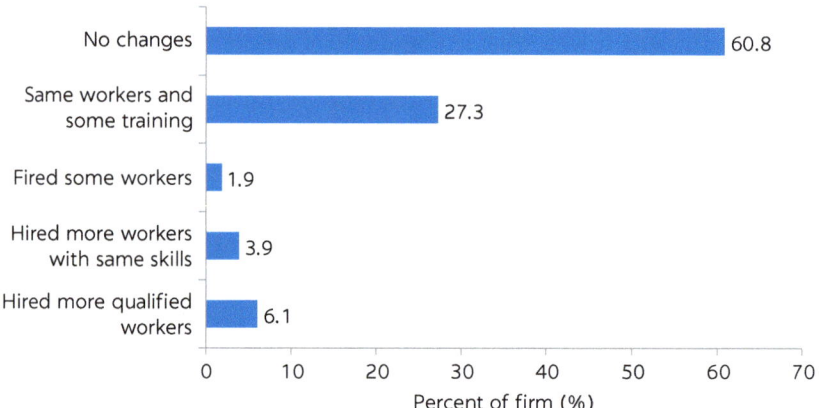

Source: Cirera et al. 2021.

FIGURE 3.14

Association between technology and job growth

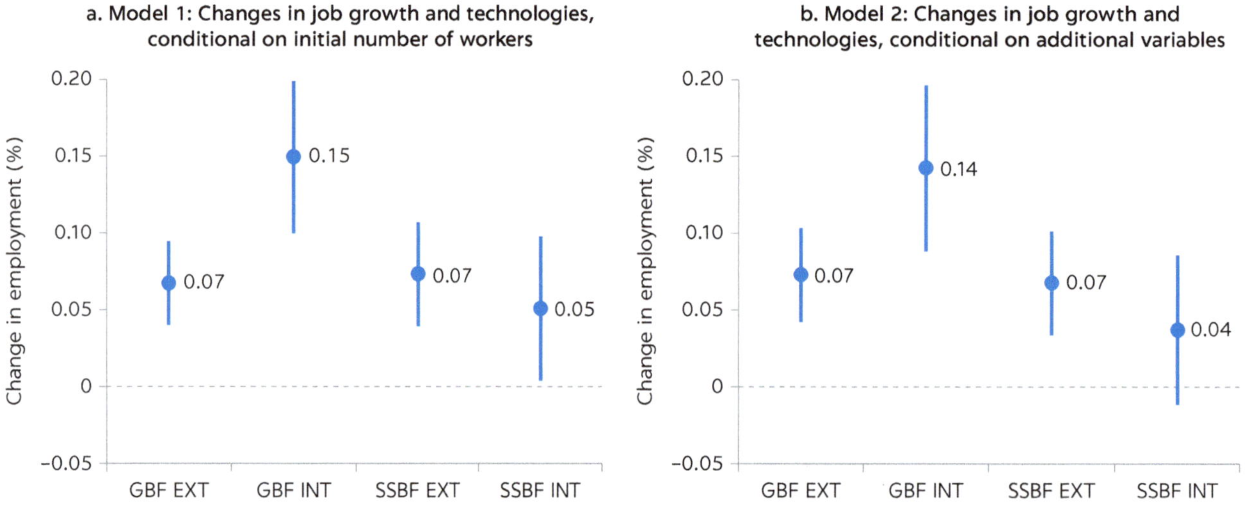

Source: Cirera et al. 2021.
Note: Point estimates at 5 percent confidence intervals. Ordinary least squares results are presented for the association between changes in the number of jobs between 2018 and 2016 and technology measures. The coefficients refer to technology measures for the GBFs and SSBFs in the extensive (EXT) and intensive (INT) margins. Models 1 and 2 control for log of initial number of workers in 2016. Model 2 also controls for firm age groups, sector, region, multinational, and exporting status.

and less precisely estimated for the intensive margin of SSBF. Although these results do not imply a causal relationship, they are in line with other findings in the literature suggesting that firms with better technologies tend to be more productive and benefit from opportunities to expand production and jobs. The correlation between firms' growth and the level of technology is more robust for GBFs at the intensive margin.

Within general business functions, firms with faster job growth tend to use more advanced technologies to perform internal-to-the firm tasks. When the association between technology adoption and job growth for specific GBFs at the intensive margin of use is analyzed, a positive and statistically significant association is found for all business functions. Interestingly, the uses of internal-to-the-firm DTs for business administration and production planning have an association with higher average job growth than the uses of external-to-the-firm DTs for upstream sourcing and downstream marketing, sales, and payment methods (figure 3.15). This finding could be partly explained by the latter requiring a more widespread and effective ecosystem of adoption, including other upstream and downstream firms and individuals as users.

Moreover, adoption of more sophisticated technologies is associated with a disproportional increase in production and service workers compared with high-skill occupations. To investigate this relationship, the correlation between the technology index and changes in the skill composition of the firm based on existing occupations in 2016 and 2018 is analyzed. The proxy for high-skill intensity is the share of high-skilled (CEOs and managers, professionals, and technicians) staff to total workers, which also includes low-skilled (clerks, production, and service workers) occupations. The differences of this share between 2016 and 2018 are used as a dependent variable. Figure 3.16, panel a, shows a negative association between changes in the skill intensity and the level of technology, controlling for the initial size of the firm. The results are statistically significant at 95 percent confidence for the GBF intensive margin and are consistent if controlling for firm age groups, sector, region, multinational, and exporting status

FIGURE 3.15

Association between intensive use of GBFs and job growth

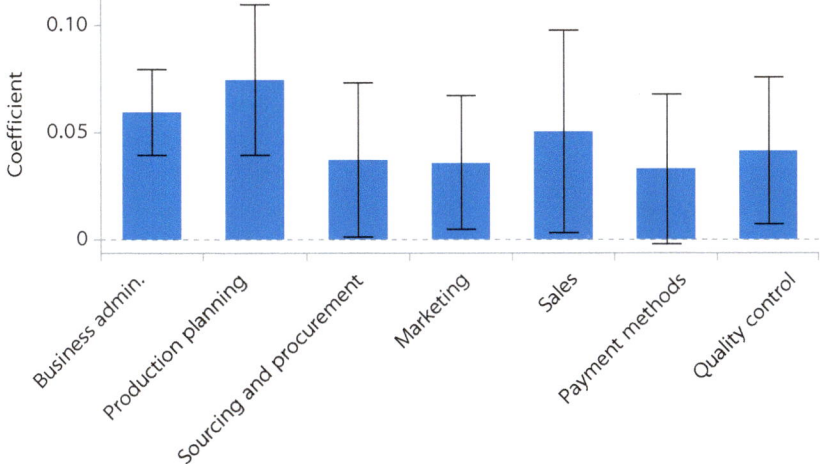

Source: Cirera et al. 2021.
Note: The figure provides the coefficients and 95 percent confidence intervals from regressions. Job growth is regressed on each specific general business function at the intensive margin, while controlling for sector, size, and regions.

FIGURE 3.16

Change in the share of high-skill occupations and technology adoption

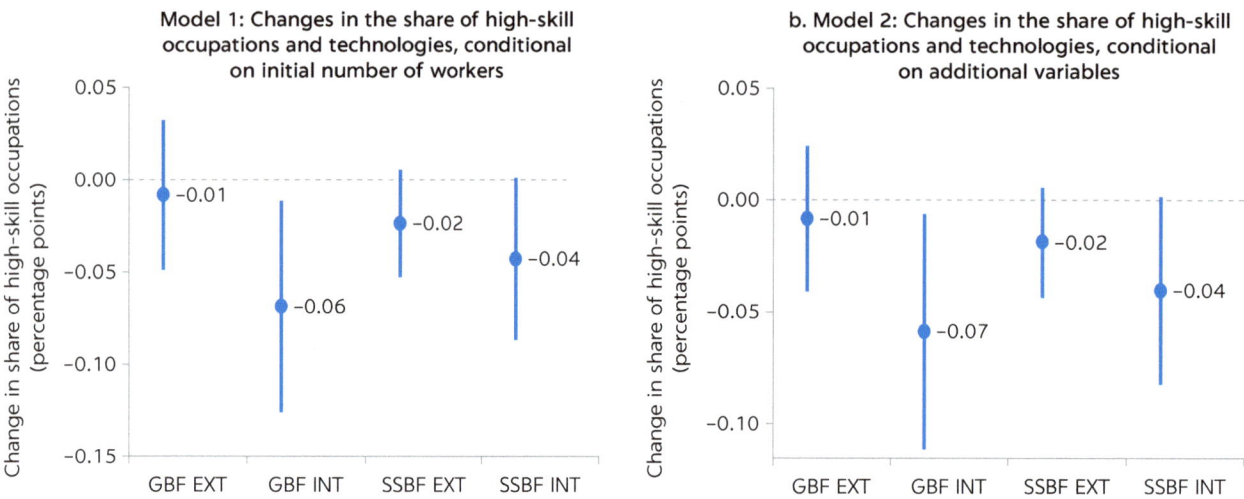

Source: Cirera et al. 2021.
Note: Point estimates at 5 percent confidence intervals. Ordinary least squares results are presented for the association between changes in the share of high-skill occupations between 2018 and 2016 and technology measures. The coefficients refer to technology measures for the GBFs and SSBFs in the extensive (EXT) and intensive (INT) margins. Models 1 and 2 control for log of initial number of workers in 2016. Model 2 also controls for firm age groups, sector, region, multinational, and exporting status.

(figure 3.16, panel b). This association does not imply a causal relationship between technology and skill intensity, but it suggests that on average firms with more sophisticated levels of technologies also generate more jobs and are more likely to increase the share of unskilled workers in their payroll.

Micro-size informal firms are also an important source of better and more jobs

Insights can be gained into adoption patterns and jobs-related benefits of DTs for micro informal firms by examining the responses to the After Access Business Survey by Research ICT Africa (RIA).[25] The RIA business survey covers a nationally representative sample of micro firms in nine countries in Sub-Saharan Africa (SSA).[26] Of the 517 firms surveyed in Senegal, more than 95 percent are micro, employing 5 or fewer full-time paid employees: 52 percent of firms are self-employed firms with no employees (of which 22 percent have unpaid family members, so can justifiably be called household enterprises), 33 percent employ one or two full-time employees, a further 11 percent employ three to five full-time employees, and only 4 percent of firms (or 20 firms) employ more than five employees. More than 90 percent are informal in the sense that they do not have all four indicators of formality used in this survey, namely, being registered with any local authority or municipality, being registered with the national revenue authority, paying local or municipal taxes (tax stamps), and being registered for the national value-added tax (VAT) or sales tax: more specifically, 56 percent are fully informal (having none of these indicators), 37 percent are semiformal (one to three indicators), and 7 percent are formal (all four indicators).[27] Compared with firms in the FAT survey, the RIA firms are much smaller: FAT covers firms with at least 5 employees and the average number of employees is 29, but the RIA survey firms' average number of employees is 1. The sectoral composition is also very different. Although 38 percent of FAT firms are in manufacturing (with a focus on food processing and wearing apparel), 57 percent of micro informal firms are in trade (primarily retail, also wholesale). Micro informal firms are also younger: the average age of FAT firms is 17, but it is 8 in RIA (the average age is 10 for the rest of SSA).

Micro informal enterprises lag even further in DT adoption, which conversely means that there is greater potential for technology upgrading and continual learning, productivity, and sales (including exporting), as well as jobs increases, for those firms able to jump the quality hurdle and join the modern economy. Though 30 percent of larger firms on average (and 28 percent of larger informal ones) use smartphones, only 18 percent of micro informal enterprises do so. Fewer than 6 percent of these firms ever had a loan (5.7 percent); however, 24 percent receive suppliers' credit (Atiyas and Dutz 2021, table 6). On average, across all micro informal firms, about 6 percent use more specialized GBF DTs, such as accounting (6.8 percent of firms) and inventory control/point-of-sale (POS) software (5.3 percent); the latter facilitates documenting and tracking the changing levels of inventories and customer purchases over time, the lifeblood of small companies, rather than an owner or manager writing them on pieces of paper and not being able to consider what they mean for company profits and growth—and is used as a proxy for better management practices among micro informal firms. It is striking that more than 27 percent of firms owned by young women use a smartphone, more than 12 percent use inventory control/POS software, and 24 percent use the internet to better understand their customers for marketing and sales—the highest shares respectively for each of these DT uses across age and gender dimensions (figure 3.17). So there is a sizable potential for some capable informal enterprises, including businesses owned by younger women, to increase their capabilities and jump the quality hurdle to increase productivity, sales, and jobs. This potential-for-upgrading conjecture is further supported by the significantly higher share of informal firms in Senegal compared with the rest of SSA firms in the RIA data set that use DTs, such as smartphones and the internet, to understand customers or e-commerce (Atiyas and Dutz 2021, table 3).

Firms that have received a loan, have electricity, are located in urban locations, are active in services, and whose business owners have vocational training are most likely to adopt and use smartphones. Regression analysis of adoption of smartphones reveals that firms with those characteristics are all statistically, significantly associated with a higher probability of smartphone adoption. Critically, whether the firm had a loan is statistically significant and is the largest correlate of smartphone adoption. Having a loan is strongly correlated with firm size, so its inclusion likely is linked to firm size not being statistically significant as an independent explanatory variable in these adoption models. It is interesting that even though being a youth-owned business is positively associated with smartphone use on the basis of unconditional correlations (with almost twice the share of youth-owned businesses using a smartphone relative to older-owner firms, 27 percent relative to 14 percent), the age of the owner is not a statistically significant conditional correlate of smartphone adoption once additional controls are introduced. Although schooling does not have a statistically significant association with increasing the likelihood of smartphone adoption across the population, schooling is a positive inducement for women. On the other hand, vocational training increases the likelihood of adoption across the population, but its net effect on women seems closer to zero or negative.

Firms that use more specialized DTs have higher productivity and sales on average and are more likely to export. As shown in figure 3.18, those informal enterprises that use more specialized DTs to facilitate GBFs, including both internal-to-the-firm management functions as well as external-to-the-firm upstream

FIGURE 3.17

The use of digital technologies by micro informal firms, by age and gender

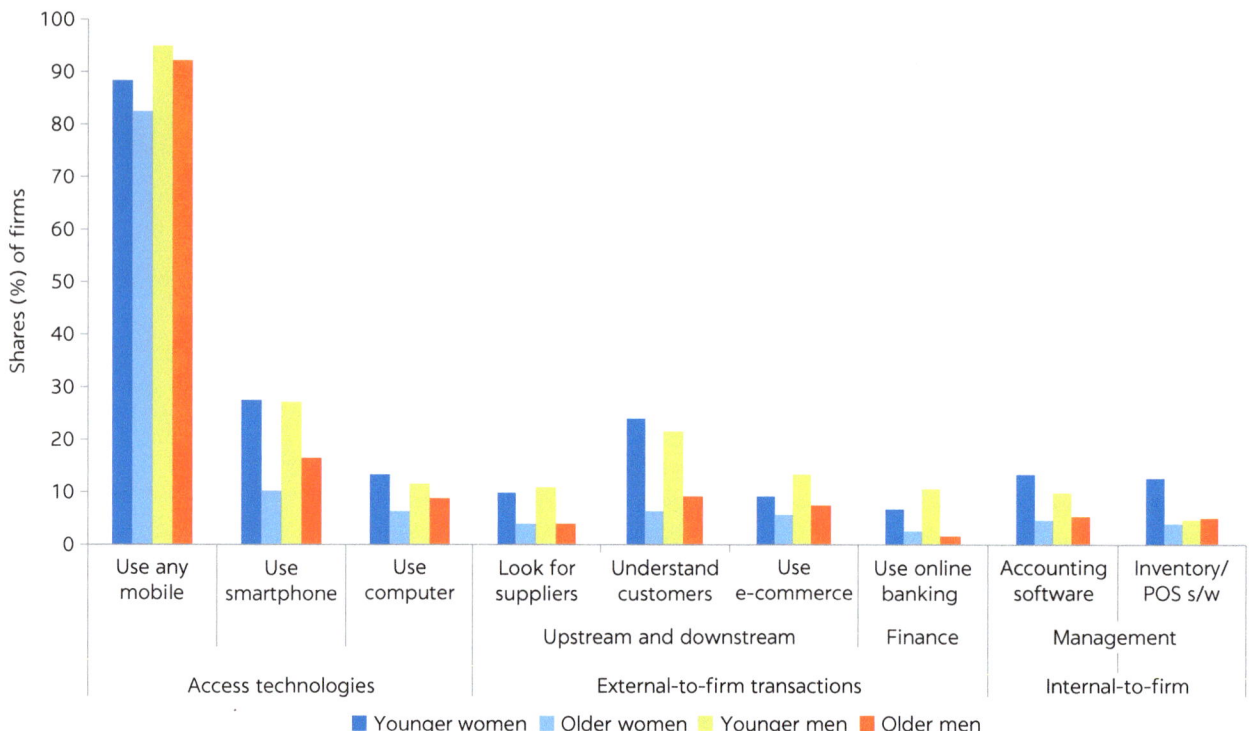

Source: Atiyas and Dutz 2021.
Note: All responses are shares (%) of firms based on weighted data. *Use any mobile* is in response to "Does the business manager have a mobile?" whether for private use, business use, or both. Smartphone users answered "yes" to "How does the business access the internet: mobile broadband (3G/4G, wireless)?" *Use computer* is a nonzero response to "How many computers does your business have?" Reported answers to "What do you use the internet for?" cover *Look for suppliers* (online), *Use e-commerce* (selling products and services online), and *Use online banking. Understand customers* is an "agree" (as opposed to "not sure" or "disagree") response to this question, "Regarding the internet/social media use, it helps to understand our customers better." The management-related questions are "Does your company use accounting software?" and "Does your company make use of inventory control/point-of-sale (POS) software?" (both asked in the computer section of the questionnaire). Appendix D.1 presents a more complete set of DTs, as well as median values, together with separate aggregates by age and gender.

transactions with suppliers and downstream transactions with customers (marketing, e-commerce, and receiving payments), have higher levels of labor productivity and total sales on average, and are more likely to export: for each DT, average values are higher for users than nonusers across all three general business outcomes. For labor productivity, the greatest difference between unconditional means for users and nonusers is for the internal-to-firm management-related DTs: users of inventory control/POS software have more than 5 times higher average labor productivity than nonusers, while users of accounting software have more than 4 times higher average labor productivity. [28]

Importantly, the businesses that have adopted and use these specialized internal-to-firm DTs also have significantly higher average labor productivity than the larger number of businesses that generically use a smartphone as an access technology. The average productivity of inventory control/POS and accounting software users is 2.2 times and 1.8 times that of smartphone users, respectively. For total sales, the top five DTs with the greatest difference between users and nonusers also include inventory control/POS and accounting software, as well as use of DTs to interact with government, recruit in the labor market, and conduct online banking. And relative to the more generic use of smartphones, the users of all five of these DTs have average sales that are at least 2.5 times higher than average. Finally, for exporting, it is striking that a relatively high share of firms reports exporting, almost 10 percent.

FIGURE 3.18

Productivity, sales, and exporting outcomes are higher for DT users relative to nonusers (unconditional outcomes)

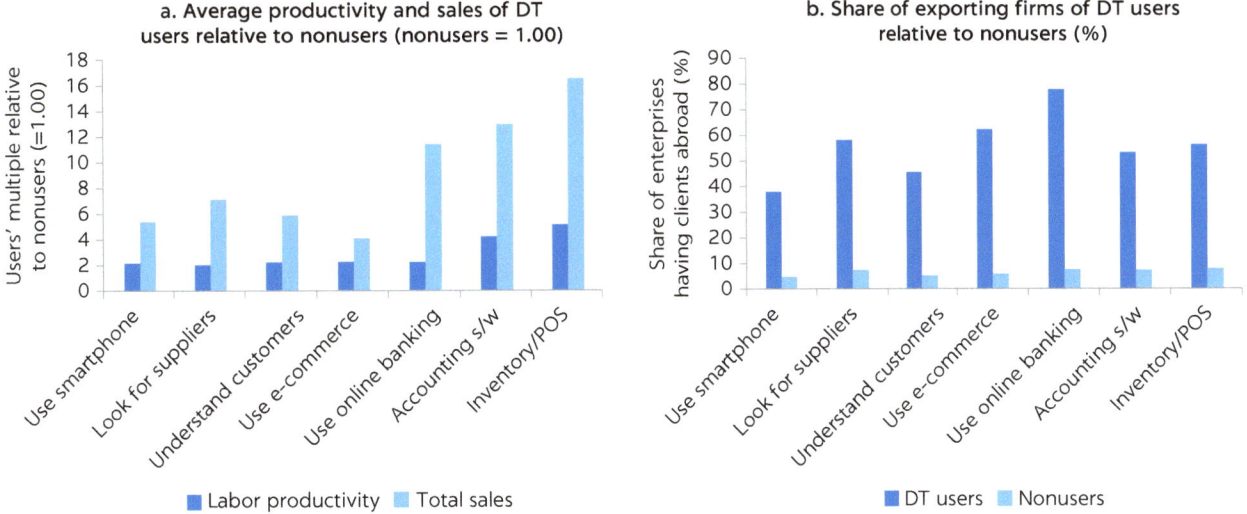

Source: Atiyas and Dutz 2021.
Note: Labor productivity is as value added (total sales minus raw materials and intermediate inputs plus water and electricity used in production) divided by total number of full-time working people, including owners. Total sales are "revenues money received by the business." Exporting reflects shares of firms that report having international customers. Labor productivity and total sales are means of monthly values, in local currency (CFAF). No use of smartphone represents use of 2G phones rather than no use of any mobile phone. Appendix D.2 presents a more complete set of DTs, as well as median values.

Roughly the same share of micro manufacturing firms exports as seen by the larger FAT firms (24 percent vs. 25 percent), with this share being actually larger for other (nontrade) services (16 percent vs. 13 percent) (Atiyas and Dutz 2021, table 4). For exporting, the top DTs with the greatest difference between users and nonusers include e-commerce and using the internet to better understand and market to customers, both uses that overcome traditional geographic distance and facilitate local producers to reach and sell to customers in other countries. Again, the businesses that have adopted and use these specialized DTs have higher exporting shares than the businesses that generically use a smartphone.

Firms that use more specialized DTs also on average generate more jobs and have higher entrepreneur profits. The RIA data allow two dimensions of job outcomes to be explored: "jobs for more people," namely, the extent to which there is a positive association between the use of specific DTs and larger firms (measured by the number of full-time paid employees plus owners), and "better jobs," namely, the extent to which there is a positive association between DTs and higher wages (averaged across all full-time employees who are paid a salary or wage) and higher profits (per entrepreneur), as indicators of better jobs for workers and owners, respectively. As shown in figure 3.19, users of DTs are systematically more likely to generate more jobs and earn higher per-owner incomes (entrepreneur profits)—across all DT uses. For jobs (proxied by firm size), two of the greatest differences between unconditional means for users versus nonusers are again inventory control/POS and accounting software. Critically, there is a strong, established causal link in the literature between productivity (including that driven by upgraded management capabilities) and increased production and sales, and the generation of more jobs (Dutz, Almeida, and Packard 2018). There is a less strong relationship between use of all types of DTs and average wages: firms that use online banking, the internet to recruit workers and interact with government, and mobile money

FIGURE 3.19

Jobs, wages, and entrepreneur profit outcomes are higher for DT users relative to nonusers (unconditional outcomes)

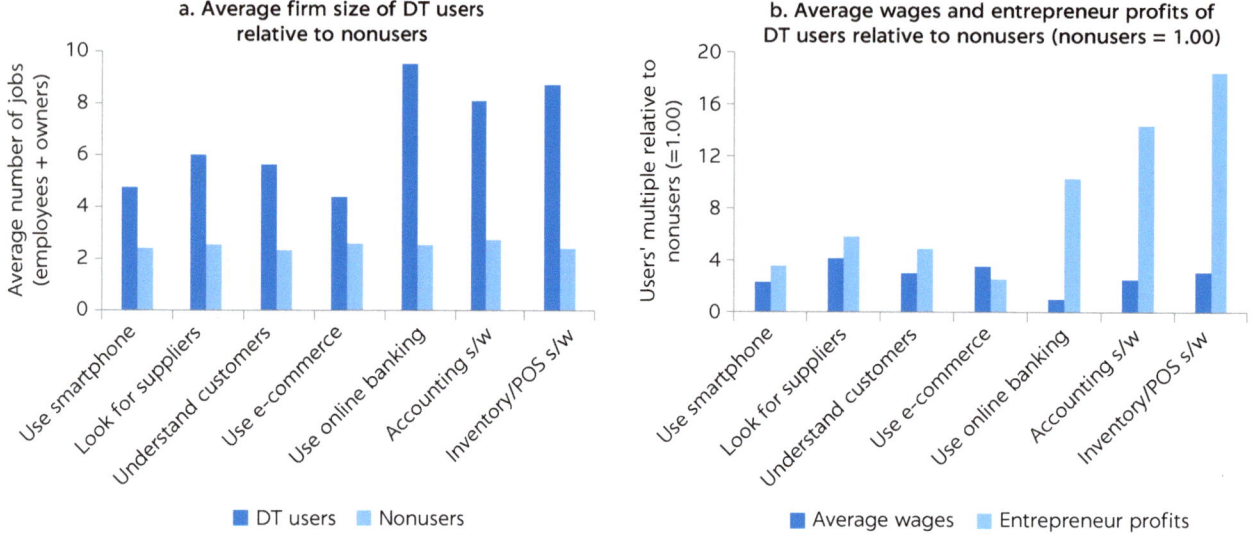

Source: Atiyas and Dutz 2021.

Note: Firm size is the number of full-time employees plus owners, averaged for each respective group. Average wages reflect salary & wages divided by full-time employees. Entrepreneurial profits are measured as value added minus salary & wages divided by number of owners. Average wages and profits are means of monthly values, in local currency (CFAF), and then normalized relative to nonusers. No use of smartphone represents use of 2G phones rather than no use of any mobile phone. Appendix D.3 presents a more complete set of DTs, as well as median values. s/w = software.

to pay taxes pay lower wages on average than those that do not use DTs (though these variables are not as directly linked to productivity as management-related internal-to-the-firm DTs). However, importantly, the use of DTs in all cases raises the average and median wages above the extreme monthly poverty line; in contrast, the median firm that does not use each of the available DTs—except for using the internet to better understand customers and to interact with government, and using accounting software—is only able to pay an average wage that is below the extreme poverty line.[29] For average profits per owner, the top DTs with the greatest difference between unconditional means for users versus nonusers again include inventory control/POS and accounting software, similar to the unconditional means for sales and for firm size. Notably, the differences between unconditional means for users versus nonusers are largest for per-owner income, relative to the other general and jobs-specific business outcomes: the average per-entrepreneur profit of users is more than 10 times the level of nonusers for each of these DTs, with the difference in average income for inventory control software users starkest at more than 18 times the income level for nonusers. Again, the businesses that have adopted and use these specialized DTs have higher per-owner incomes than the businesses that generically use a smartphone.

Importantly, the most significant conditional correlates of productivity, sales, and jobs are internal-to-the-firm general business functions, namely, inventory control/POS software as a management tool, in addition to having a loan and electricity, and the owner having vocational training. Controlling for the effects of having a loan, having access to electricity, firm sector and size, and age and gender dimensions, the use of inventory control/POS software as a proxy for internal-to-the-firm management-related DTs is the only consistently significant conditional correlate of productivity, sales, and jobs—in addition to having a loan and having electricity (figure 3.20). This finding suggests that the adoption

FIGURE 3.20

Significant conditional correlates of productivity, sales, and jobs

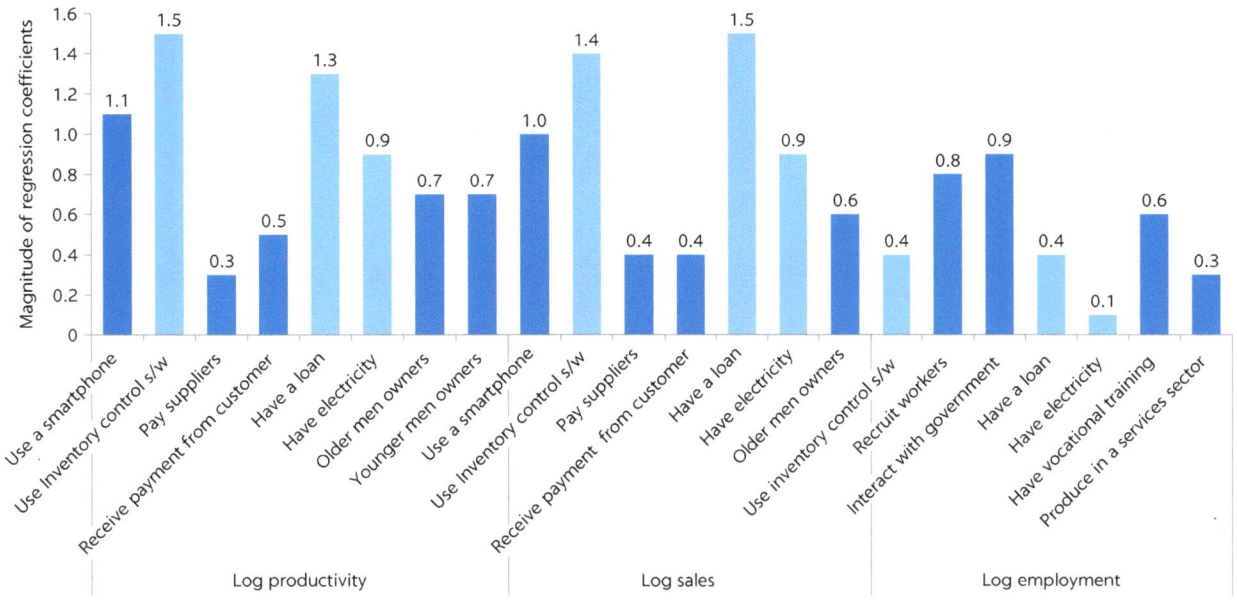

Source: Atiyas and Dutz 2021.
Note: This figure represents those conditional correlates of productivity, sales, and jobs that are significant at least at the 5 percent level across all regressions, not including economic sectors of activity. "Jobs for more people" is proxied by larger firms, measured by the log of employment, the number of full-time employees plus owners. Light blue bars highlight those variables that are similarly significant across all three business outcomes: use inventory control software, have a loan, and have electricity. s/w = software.

of relatively simple technologies to improve basic management functions, as well as having access to finance and electricity, are important components of a more inclusive jobs growth agenda.

START-UP ENTREPRENEURSHIP: TOWARD MORE FIRMS

The challenge of achieving better and more firms requires not only technology upgrading of existing firms but also an environment that supports such growth. Starting a new business, a first step in any entrepreneurial activity, requires ability and knowledge to convert ideas into goods and services and market them. This process requires (a) factor markets that provide access to basic production resources, such as labor and human capital as well as entrepreneurial characteristics and firm capabilities; (b) access to capital, with access to finance at affordable rates and instruments, and access to markets; and (c) institutions that favor this process of creation, production, and marketing and enable the flow of ideas, technology, talent, and resources (Audretsch, Cruz, and Torres 2020).

A vigorous and dynamic entrepreneurship ecosystem is critical. Many of these complementary factors face mobility costs across sectors and regions, making the local environment around the firm critical to enhancing productivity and competitivity. The availability of better resources combined with good institutions tend to facilitate the process of creating new firms, the expansion of young firms, and the technological catchup of firms overall. Yet the key factors necessary to strengthen an entrepreneurial ecosystem (for example, knowledge, human capital, entrepreneur talent, and managerial capacity) are structural and demand investment, resources, and time to mature. Thus, identifying the

FIGURE 3.21

Entrepreneurship performance

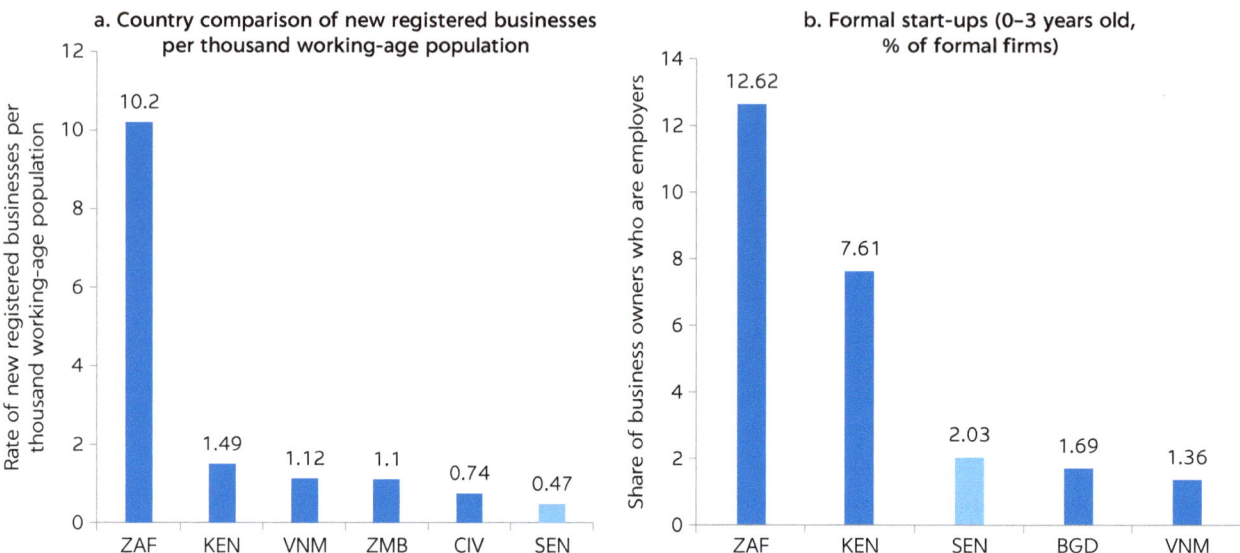

Source: Cruz, Torres, and Tran 2020. Panel a: World Bank Entrepreneurship Database (2018) or latest year available; panel b: World Bank Enterprise Survey (most recent year available).

potential of local ecosystems can help to target and deliver more customized policies that can benefit from spillover effects through economies of agglomeration.

Small, informal firms and weak dynamism characterize the landscape of the private sector in Senegal. Businesses in Senegal employ on average one worker (in addition to the business owner) and, depending on the definition, only between 3 and 13 percent of firms are formal.[30] Measures of business creation, scale-up of firms, and adoption of technology are significantly weaker compared with other regional and global peers. For example, the rate of new registered businesses is 0.47 per thousand working-age population in Senegal but 10.2 in South Africa (figure 3.21). Similarly, the fraction of formal start-ups in a representative survey comparable across countries is 2 percent in Senegal but 13 percent in South Africa. The average formal firm after 20 years in operation has only 67 employees, implying an anemic growth rate.[31]

Entrepreneurship ecosystems

Gaps in supply and demand factors in Senegal mirror the performance in entrepreneurship outcomes. Capital per worker in Senegal is about a third of the capital per worker in South Africa. Only 8 percent of adults age 25 or older have completed at least upper secondary education, whereas this figure is 65 percent in South Africa, and no university in Senegal competes in international rankings. Fewer than 9 percent of formal businesses export at least 10 percent of their sales, even if 37 percent import intermediate goods from abroad.[32] The potential cultural and regulatory barriers to entrepreneurship do not seem as binding in Senegal as in other countries, as shown by its relatively small gaps relative to its peers.

Improving these fundamentals on a broad scale, while critical for long-term growth, requires time and extensive amounts of resources. Accumulating fundamental capabilities such as stocks of human capital and knowledge is a gradual process. For example, despite steady progress in past decades, both education enrollment and attainment in Senegal remain low. The current adult literacy rate stands at 43 percent, below averages for both SSA and low- and middle-income countries, and household heads have on average only three years of schooling (World Bank 2018). Senegal will need to continue these investments. Achieving the rate of transformation envisioned in the Plan Sénégal Émergent will require growth rates higher than what is allowed by improving fundamentals alone.

Boosting strategic ecosystems could accelerate convergence in specific sectors and regions, and potentially yield higher returns through spillover effects. Traditionally, countries have tried to improve productivity by targeting sectors or regions that are presumed to produce positive technological externalities. But countries typically have lacked the empirical tools to do so systematically. Identifying how economic activities agglomerate and correlate spatially helps inform the design of such policy interventions. Investments in strategic ecosystems—specific sectors in specific regions with higher potential for spillover effects—can lead to higher gains in productivity and job growth.

A new methodology using firm-level census data was applied to identify strategic entrepreneurship ecosystems in Senegal, taking into consideration the agglomeration of firms in terms of product diversity and quality of firms. The algorithm to identify these ecosystems evaluates the diversity and the quality of geographical agglomerations of firms.[33] The indicator of diversity first looks for statistically significant agglomerations of communes with a *high density of establishments within each 4-digit subsector* in the value chain, and then counts the number of subsectors for which a commune is part of an agglomeration.[34] The indicator is then sorted into three broader measures of diversity: (a) no agglomerations, (b) agglomerations in one subsector (monosector), and (c) agglomerations in more than one subsector (multisector). Similarly, the indicator of quality first looks for agglomerations in measures of *business dynamism*—firms with more than 20 employees and young firms (0–4 years)—and measures of the *potential for additional growth*—formal firms and firms where the manager has tertiary education. The indicator of quality then counts the number of quality indicators for which a commune is part of an agglomeration and is sorted into three broader measures of quality: (a) no quality agglomerations, (b) agglomerations in one quality indicator (monoquality), and (c) agglomerations in more than one quality indicator (multiquality).

On the basis of those results, a new typology of entrepreneurship ecosystems is proposed, by grouping them as *multiquality, monoquality,* or *potential.*[35] *Multiquality* ecosystems exhibit agglomerations in more than one quality indicator and agglomerations in at least one subsector within the value chain. *Monoquality* ecosystems exhibit agglomerations in one quality indicator and at least one subsector. *Potential* ecosystems exhibit agglomerations in more than one subsector but no quality agglomerations. Sales per worker and sales per firm across ecosystems serve as tests of the algorithm—these proxies for productivity were not used as indicators to measure quality, but in general they increase with the quality of the ecosystem. All indicators of agglomeration for this analysis, including quality indicators, are relative to firms in Senegal. Although the quality for the ecosystem is based on diversification in terms of a quality indicator, the mono- versus multisector characteristics show how specialized or diversified this ecosystem is within an aggregated sector. This information might be relevant to defining policy strategies associated with information and knowledge, such as technology extension programs. Table 3.5 shows that high-potential

TABLE 3.5 **Productivity and sales in entrepreneurship ecosystems**

SECTOR	AVERAGE SALES PER WORKER (2016 CFAF MILLION)			AVERAGE SALES PER FIRM (2016 CFAF MILLION)		
	MULTIQUALITY ECOSYSTEMS	MONOQUALITY ECOSYSTEMS	POTENTIAL ECOSYSTEMS	MULTIQUALITY ECOSYSTEMS	MONOQUALITY ECOSYSTEMS	POTENTIAL ECOSYSTEMS
Agribusiness	16.58	3.38	2.12	68.41	12.88	4.21
Manufacturing	23.33	2.74	4.57	57.77	5.15	8.37
Services	14.11	5.37	3.11	55.25	9.73	6.57
Retail	29.30	14.26	7.91	36.71	16.17	9.15
Tourism	11.52	3.33	2.31	35.42	5.18	4.66
Digital	75.60	3.18	1.91	286.76	4.74	2.82

Source: Cruz, Torres, and Tran 2020.

Note: Agribusiness includes firms in agriculture and food processing, *Manufacturing* includes firms other than food processing, and *Services* includes firms other than *Retail* and *Tourism. Multiquality ecosystems* exhibit agglomerations in more than one quality indicator and agglomerations in at least one subsector within the value chain. *Monoquality ecosystems* exhibit agglomerations in one quality indicator and at least one subsector. *Potential ecosystems* exhibit agglomerations in more than one subsector, but no quality agglomerations.

ecosystems have higher productivity per worker and also higher sales per firm, compared with maturing and incipient ecosystems.

In agribusiness, multiquality ecosystems are agglomerated in Dakar and southern Casamance. Casamance and Dakar are two entrepreneurship ecosystems in agribusiness with high densities of businesses in several subsectors within agribusiness, combined with spatial agglomerations of high-quality firms relative to other regions in Senegal (figure 3.22). These agglomerations are highly diverse, and businesses in these regions exhibit high dynamism and the potential for more growth. They account for 2.3 percent of plants and 4.9 percent of employment in the 2016 establishment census. The Niayes-North and Saint-Louis regions are monoquality ecosystems: agglomerations in these regions are diverse, but with quality agglomerations in only one indicator. These communes account for 2.8 percent of establishments and 5.4 percent of employment in the census.

Diversity and quality in other value chains, including the digital economy, are mainly agglomerated in Dakar and Diourbel. In tourism, the east of Ziguinchor is a potential agglomeration, whereas Dakar and the south of Thiès are densities with multiquality and high potential. Retail, manufacturing (other than food processing), services (excluding retail and tourism), and tourism account for the largest shares of employment in the 2016 establishment census (in addition to agribusiness).[36] In retail, which in general accounts for a significant fraction of jobs created, geographical agglomerations with high potential correspond to population agglomerations—very few communes outside Dakar, Thiès, and Diourbel exhibit significant establishment densities. Agglomerations in services and manufacturing are scattered across Senegal, but there are no ecosystems with high potential in services, and only one in manufacturing (Dakar). In manufacturing, Kolda, Tambacounda, and Kédougou exhibit agglomerations in a diversity of subsectors, but no statistically significant agglomerations in indicators of quality. The analysis suggests a challenge in regional inequality and the importance of addressing challenges faced by maturing and incipient ecosystems as well, if the policy objective is to reduce regional disparities.

Boosting entrepreneurship in high-potential (mono- and multiquality) ecosystems could significantly increase employment by leveraging on the spatial

FIGURE 3.22

Local agribusiness ecosystems in Senegal

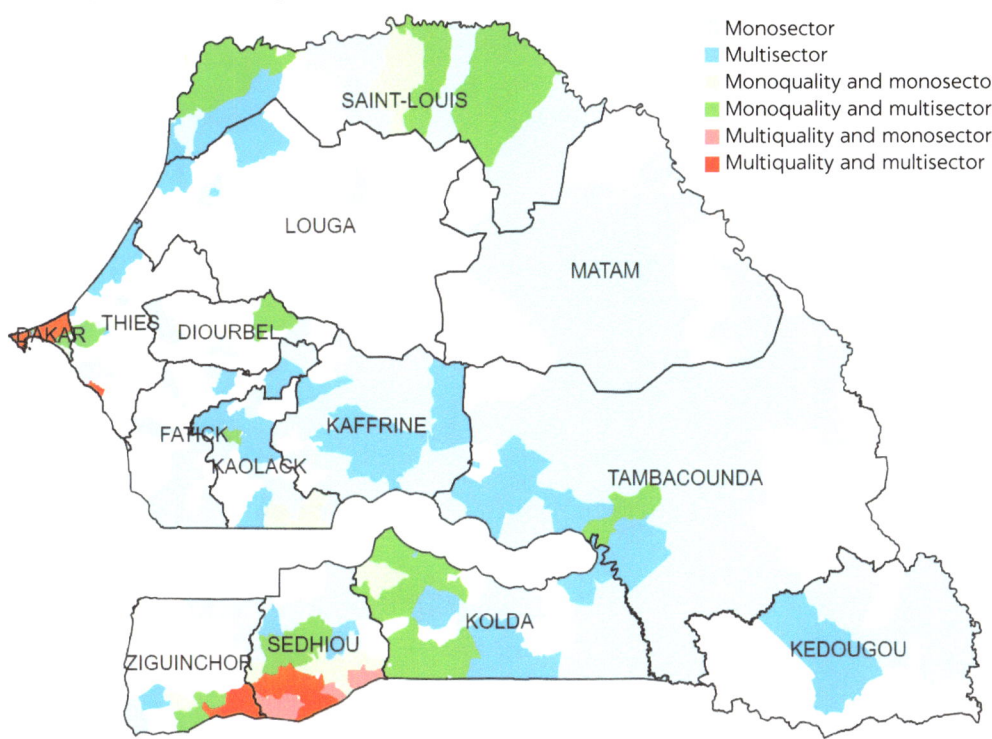

Source: Cruz, Torres, and Tran 2020.
Note: The analysis is based on indicators at the commune level generated through microdata from the Recensement Général des Entreprises (RGE), ANSD.

contagion of positive shocks. The spatial correlation in economic activity could amplify the effect of interventions targeted to strategic entrepreneurship ecosystems. For example, boosting entrepreneurship only in high-potential communes in agribusiness in Dakar, Casamance, and the Niayes-North regions (modeled as an exogenous increase in plants and sales in figure 3.23) affects employment in these communes (direct effect), and could also affect Thiès, the rest of Ziguinchor, Sédhiou, and Kolda, and extensive regions of Louga and Matam through spillover effects. This analysis exploits a spatial regression of log employment on log plants and log sales in agribusiness across communes in Senegal. The weighting matrix in the spatial regression assigns a value of 1 to neighboring communes that share a border and 0 otherwise. In the exercise, plants and sales in high-potential and maturing communes are increased by 10 percent (increase of 4.8 percent as a fraction of total number of plants in agribusiness) and the resulting increase in employment is 11 percent (as a fraction of total employment in agribusiness).

Main barriers to boosting entrepreneurship

Agribusinesses in multiquality ecosystems in Senegal face barriers that significantly differ from the constraints of the typical firm. The top constraint for businesses in Senegal, as reported in the establishment census, is the difficulty of distributing and selling final goods. About 30 percent of firms report this barrier, which is especially stringent in multiquality ecosystems

FIGURE 3.23

Spillover effects from interventions that boost entrepreneurship in mono- and multiquality ecosystems in agribusiness

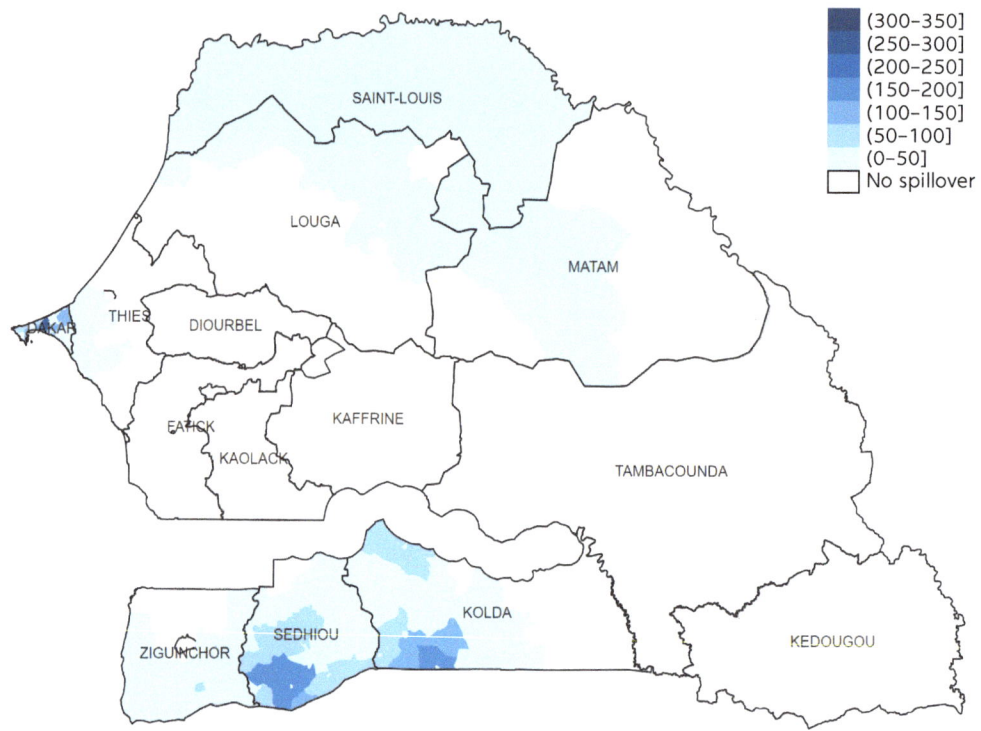

Source: Cruz, Torres, and Tran 2020.
Note: The exercise exogenously increases establishments and sales in high-potential and maturing communes in Dakar, Casamance, and the Niayes-North region, and then predicts the change in employment not only in the targeted communes but also in neighboring regions. Darker areas in the map identify regions more affected by increasing number of jobs when boosting entrepreneurship in mono- and multiquality ecosystems. The analysis is based on indicators at the commune level generated through microdata from the Recensement Général des Entreprises (RGE), ANSD.

for tourism (32 percent), retail (36 percent), and manufacturing (35 percent). This barrier is followed by lack of suitable premises and high taxes. In agribusiness, lack of specialized technology and an inadequate or costly labor force are relatively more stringent for high-potential ecosystems. In tourism, another sector that generates a significant number of jobs in Senegal, the two factors that deviate from the average, in main perceived barriers, are access to markets and regulations (figure 3.24). Better access to technology, access to finance, and better regulations are also statistically associated with entrepreneurship performance (Audretsch, Cruz, and Torres 2020).

Removing barriers in strategic communes could significantly boost entrepreneurship through direct and spillover effects. Estimates from a spatial regression of entrepreneurship outcomes on constraints (and other controls) using aggregates at the commune level suggest that the potential positive spillover or contagion of effects from boosting entrepreneurship in specific communes is significant. Targeting the lack of qualified labor to relax the constraint by 1 percent, for example, could increase average sales per worker across communes in Senegal by 0.541 percent. A majority of this effect (65 percent) comes solely

FIGURE 3.24
Deviation from the average perceived barriers in multiquality entrepreneurship ecosystems

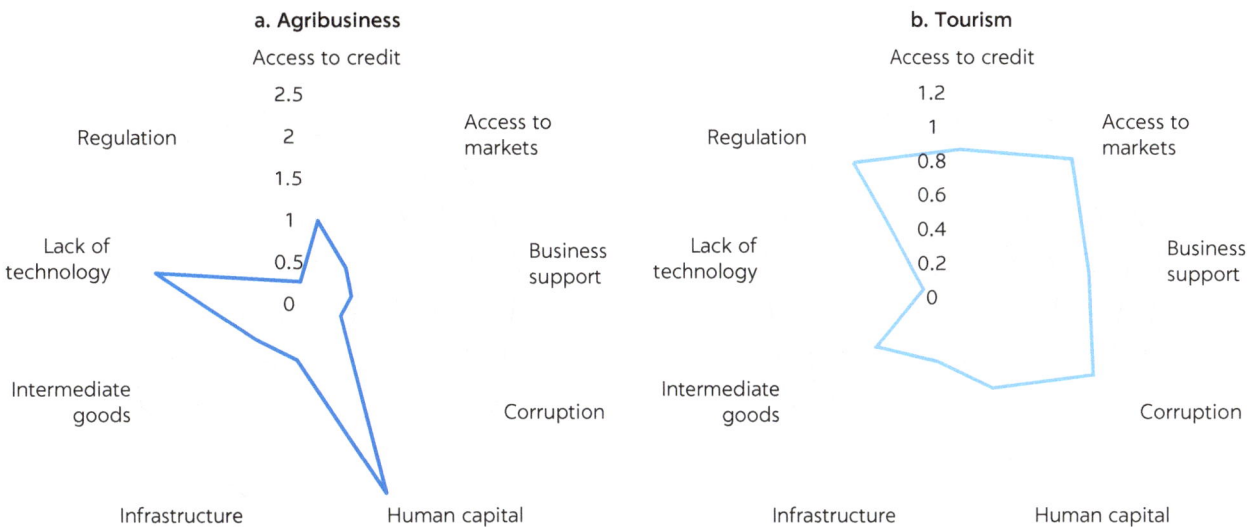

Source: Cruz, Torres, and Tran 2020.

from boosting entrepreneurship in neighboring communes (Cruz, Torres, and Tran 2020).

A stronger digital entrepreneurship ecosystem could also support other businesses by providing digital solutions that address some of these barriers. Digitization for Agriculture (D4Ag) is a national priority for Senegal (SN2025) strategy. With about 40 D4Ag tools identified, Senegal has an emerging ecosystem it can build on. However, the country lags behind leaders such as Kenya (which accounts for 132 D4Ag tools), Nigeria (88 tools), Tanzania (82 tools), Uganda (80 tools), Ghana (60), and Rwanda (46 tools). A good example of this potential solution is CommAgri, a digital application aiming at improving access to markets for farmers (box 3.3). Intermediary support organizations are likely important to strengthen digital entrepreneurship ecosystems, but face their own challenges (box 3.4).

BOX 3.3

Digitization for Agriculture (D4Ag) in Senegal: CommAgri and Commango case studies

CommAgri, the solution used to manage the *Feed the Future Senegal Nataal Mbaye* agricultural extension program, aims to improve cereal market systems in Senegal. The CommAgri platform, financed by the US Agency for International Development, helps farmers' collectives more effectively manage relationships with their members, who are mostly smallholder farmers. It also helps them manage purchases of inputs and sales

continued

Box 3.3, *continued*

of crops as well as loan provision and repayments. It facilitates both agricultural extension and the collection of farm-level acreage and production data. According to a recent evaluation, the introduction of the tool led to a 161 percent increase in maize yields and a 73 percent increase in millet yields for 25,000 farmers across the country. The tool also accelerated the loan approval process during the rainy season, allowing some farmers to plant two seasons of rice. In 2016, the total for rainy-season loans reached US$12.1 million. This experience has inspired the development of a similar app, called Commango, which aims to better connect mango producer collectives with markets.

The Commango app is still under development, supported by an IFC Advisory project implemented with APIX, the Senegalese investment promotion agency. It was developed to support the commercialization of mangoes in Casamance, a region that accounts for 46 percent of Senegal's mango

production. However, the region exports only 5 percent of its production. Isolated producers face serious phytosanitary, logistics, and commercialization challenges. Potential buyers and financial intermediaries do not have detailed and reliable information on production. The app was developed to help bridge these gaps between producers, off-takers, and financiers. It already includes more than 12,000 mango producers. Although its current focus is limited to collecting, managing, and sharing production information (owner, location, quantities, varieties, and so on), upcoming developments will aim at enabling transactions (selling/buying), as well as connecting it to financial intermediaries with the capacity to fund harvest campaigns. Partnerships with other platforms providing additional services to mango producers will also be explored to identify an appropriate business model that aims at financial sustainability of the app.

Source: Dalberg 2020.

BOX 3.4

Digital entrepreneurship ecosystems in Senegal

The high-potential (multiquality) digital entrepreneurship ecosystem in Senegal is concentrated in Dakar. There are some additional clusters in Diourbel and Kaolack (figure B3.4.1a). According to a recent survey conducted by the Ministry of Digital Economy of Senegal, most firms that are producing digital content are young (start-ups) and focus on providing digital solutions related to e-commerce and logistics, followed by public administration and social media (figure B3.4.1b). On average, these digital firms have managers with higher levels of education and higher sales per workers.

Senegal seems to be growing rapidly in intermediary institutions that support digital entrepreneurship. Results from a recent survey by the World Bank of incubators, accelerators, and other organizations that support entrepreneurship in Senegal suggest that the emergence of these structures has increased in speed over the past few years. Most organizations started their activities after 2010 (figures B3.4.1b and B3.4.1c).

Supporting organizations in Senegal aim to play a key role in digital ecosystems by developing a collaborative mind-set. In a survey, they responded that they mainly offer networking opportunities,

continued

Box 3.4, *continued*

FIGURE B3.4.1

Digital ecosystems in Senegal: Firms and intermediary organizations

a. Digital ecosystems

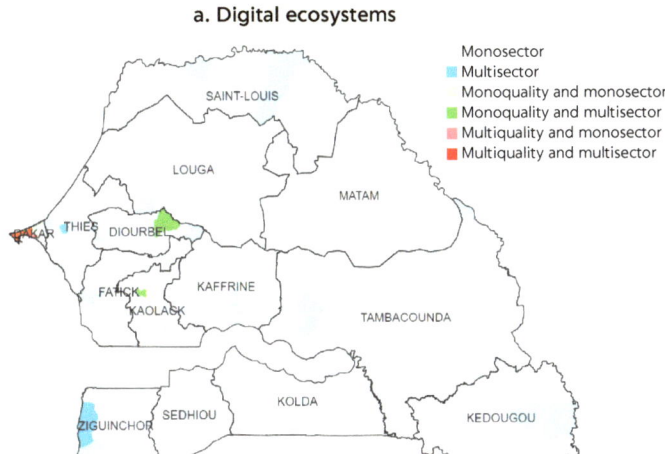

Monosector
Multisector
Monoquality and monosector
Monoquality and multisector
Multiquality and monosector
Multiquality and multisector

b. Type of intermediary organization

c. Services provided

Sources: Cruz, Torres, and Tran 2020; World Bank Mapping of Intermediary Organization in Senegal 2020; Ministry of Digital Economy of Senegal.

a workspace, technology extension services, and programs to increase managerial capabilities (figure B3.4.2). They report serving businesses across Senegal—only around 40 percent are focused on Dakar—and most have no sectoral orientation. The digital ecosystems in Senegal are becoming more diverse and interconnected. Thirty-eight percent of managers of supporting organizations are women. Most top managers helped start or owned businesses before. About 95 percent of these organizations market to individuals, followed by 65 percent that market to firms. The most common methods for recruiting new customers are "online advertising" (70 percent) and "word of mouth" (60 percent). Most organizations (63 percent) are not oriented toward specific sectors.

The top objective of these organizations is to build up the ecosystem, but they seem to face their own

continued

Box 3.4, *continued*

FIGURE B3.4.2
Barriers and gaps

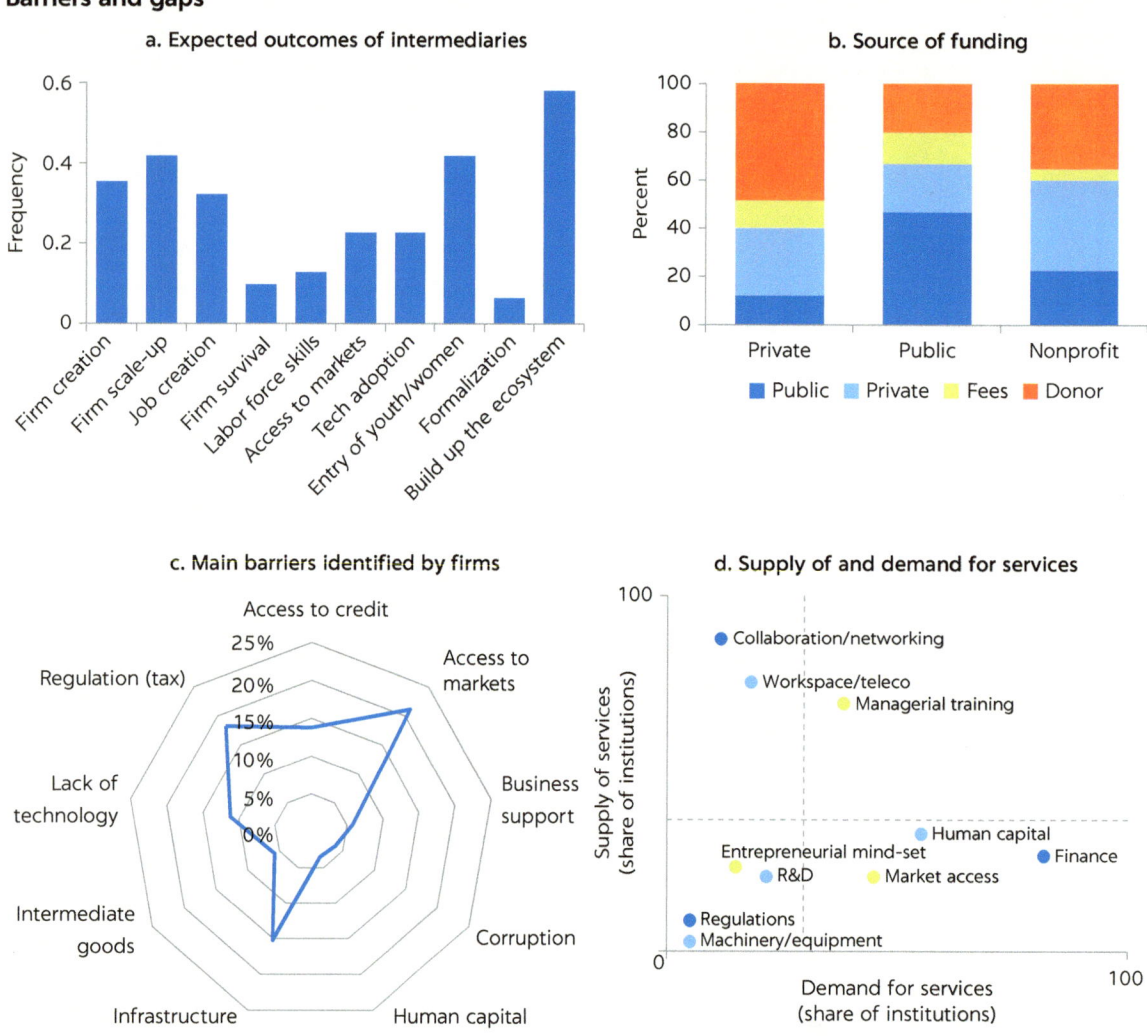

Sources: World Bank Mapping of Intermediary Organization in Senegal 2020; Ministry of Digital Economy of Senegal.

challenges, including the lack of rigorous monitoring and evaluation systems. Adequate and competent support mechanisms—both public and private—are critical to boosting firm creation, firm scale-up, and innovation. However, support mechanisms for digital entrepreneurship in Senegal are still in a nascent stage. A significant fraction of their budget comes from donors, and they often face difficulties in securing financing. In addition, at least 40 percent of their budget is spent on administrative costs and other operating expenses.

An important challenge for building a stronger entrepreneurship ecosystem in Senegal is to improve the quality of supporting mechanisms and match supply and demand. Although access to finance, access to markets, and human capital are subjectively reported as important barriers for digital firms, a larger share of services is focusing on building collaboration and networking. Improving the matching between main challenges identified by firms and other key actors and the supply of services with high quality are key.

continued

Box 3.4, *continued*

The allocation of future resources for the more technologically sophisticated programs should benefit from a screening based on business priorities in high-potential ecosystems (for example, designing solutions to solve problems in specific agribusiness value chains). An important feature of digital entrepreneurship ecosystems in Senegal is that a significant number of resources used by either private or public institutions are coming from donors or the public sector. The main barrier identified by digital firms in high-potential ecosystems is associated with "difficulties in distribution and sales of product." Identifying potential opportunities to address barriers to meet the needs of other businesses could lead to important spillover effects toward the entire economy.

TECHNOLOGY AND ENTREPRENEURSHIP POLICIES FOR BETTER AND MORE FIRMS

A key priority for Senegal is to promote technological upgrading of existing firms and facilitate the creation of more firms of better quality. The analysis of obstacles for technology adoption and entrepreneurship has identified some common challenges to generating better and more firms. Among the common barriers impeding the adoption of better technologies and the growth of high-potential entrepreneurship ecosystems, five strategic areas of policy priorities have been identified: (a) access to information and knowledge (firm capabilities, including management and worker skills); (b) access to markets; (c) regulations to promote entry and innovation; (d) access to finance; and (e) digital solutions across these areas to support better and more firms. This section proposes policy interventions and discusses how business-oriented digital technologies can be used to help address these challenges.

Senegalese government bodies providing business support would benefit from improved priority-setting and stronger coordination across different programs. The Agence de Développement et d'Encadrement des Petites et Moyennes Entreprises (ADEPME), the Bureau de Mise à Niveau du Senegal (BMN), and the Délégation générale à l'Entreprenariat Rapide des Femmes et des Jeunes (DER), among other local institutions, are important in entrepreneurship and private sector development support. The Ministry of Economy, through the Direction du Développement du Secteur Privé, could have an important role in strengthening the coordination of these programs and establishing a systematic monitoring and evaluation (M&E) process for firms that benefit from these initiatives. Recent experiences with the implementation of Public Expenditure Reviews (PERs) to assess public programs supporting entrepreneurship, technology adoption, and frontier innovation, led by the World Bank, suggest that countries often face significant challenges in adjusting their policy mix toward priority development strategies. Senegal has been expanding its efforts to support business in recent years. However, data on the resources allocated to these initiatives and their effectiveness are still limited. Establishing procedures such as a logical framework for all programs, a prioritization consistent with budget allocation, and an effective and continual M&E system could enhance government capacity to implement and adjust these programs.[37]

One way to enhance public policy impact would be to anchor government support programs in vertical industry value chains where public-private coordination problems can be solved through focused, industry-specific working groups, supported by a punctual execution plan and an effective delivery unit. These vertical industry value chains, such as for specific horticulture products, would focus on solving coordination problems that arise. Senegal's public support for businesses—for technology extension and firm capabilities, for access to domestic and export markets, for entry and innovation promotion, for access to finance, and for the development and scaling up of digital solutions across these areas—could benefit from being anchored in these value chains. This is because many of the problems faced by firms are specific to their value chain. Efficient production by firms often requires highly specific public inputs, such as industry-specific laws, regulations and permits, industry-specific skills, work practices, quality standards and accreditation, industry- and location-specific infrastructure, and associated financing. Such public inputs are a form of positive coordination externality, benefiting all firms in the industry. Inputs are often required in bundles specific to the value chain and to the needs of specific types of producers, such as a combination of technology, skills training, financing, insurance, and access to markets for micro farms in specific horticulture value chains.[38] These inputs are typically underprovided in private markets absent government intervention in the broader public interest. Their absence can prevent productivity growth, while their presence can enhance it. The case of Peru's "mesas ejecutivas" (MEs) provides a persuasive case for how such public-private coordination problems can be solved. MEs are public-private working groups to identify and remove specific bottlenecks and add missing public inputs. In the period between December 2014 and May 2016, the Ministry of Production of Peru created eight MEs: six industry-specific or vertical (forestry, aquaculture, creative industries, textile, gastronomical, and agro-exports, in this order over time) and two cross-industry or horizontal (logistics and high-impact entrepreneurship). The government helped address specific identified public inputs, such as help to comply with technical requirements for new export markets or to design an industry-specific training program—but did not provide tax exemptions or subsidized credit. As an example, the forestry ME obtained coordination between different public entities across line ministries and different levels of government to solve specific, jointly identified, public good–type bottlenecks. Achievements included a new law and regulation recognizing plantation trees as crops, removing the requirement of a permit to extract wood from plantations, and reducing the registration time of plantation properties from up to one year to three days. A new protocol with the same timber resource standard at national, subnational, and local government levels was passed. Investors and reforestation companies started a process to establish, for the first time, a business association that represents their interest. And some of the largest global forest funds started to invest in Peru.[39] For effective implementation of such support, a time-bound action plan that is agreed upon within the value chain can be decisive, with clear goals that can be tracked on an easy-to-monitor dashboard by an effective delivery unit.[40]

Technology extension and firm capabilities programs

Improving access to information and knowledge is essential to enhancing firm capabilities and facilitating technology adoption. Empirical evidence suggests that improving managerial capabilities and the organization of firms can have sizable and durable effects on firm productivity.[41] An experiment in India showed significant causal effects from management consulting programs that provided recommendations for improving management practices to build firm capabilities. These effects were persistent 10 years later (Bloom et al. 2013, 2018). Another experiment in Mexico showed the benefits of management consulting services on total factor productivity and return on assets, including a persistently large increase (about 50 percent) in the number of employees and the total wage bill, even five years after the program was completed (Bruhn, Karlan, and Schoar 2018). In Pakistan, randomized experiments with soccer ball producers suggest that providing the right incentives to workers is key to facilitating technology adoption with positive impact on firm performance (Atkin et al. 2017). In Brazil, a program that provided coaching and consulting on management and production practices led to changes in the organization of firms that facilitated exports for smaller firms (Cruz, Bussolo, and Iacovone 2018). A key question regarding scaling up management consulting programs is related to the fact that these programs are usually costly. Yet a recent experiment in Colombia suggests that a group-based approach could be an effective option (Iacovone, Maloney, and McKenzie 2019). Digital technology tools such as inventory control and point-of-sale software and accounting software could also be low-cost solutions to allow managers to understand their own data and develop better business planning capabilities over time.

Programs aiming to support technology extension and firm capabilities could initially target high-potential (mono- and multiquality) ecosystems that benefit from economies of scale, agglomeration, and potential spillovers, and experiment with the use of greater DTs in supporting the program. A more technologically sophisticated program should be focused on firms with more sophisticated capabilities and needs. A customized part of the program should be offered to enterprises with less sophisticated capabilities, including informal farms. In this type of program, technology support policies could help informal farms jump the quality hurdle to integrate into formal, more-structured value chains. Based on the experience of the productive alliances model in agriculture, a four-step approach could be considered that facilitates links and learning between larger and smaller enterprises: (a) organizing farmer networks into associations or cooperatives; (b) linking these farmer networks with markets by supporting contract models, negotiations, and oversight processes, together with digital payment systems; (c) providing technical assistance to meet market quality requirements through technology adoption extension services, strengthened quality control systems and certification, and support for simple upgrading of GBFs with a focus on inventory control/POS solutions, accounting, and other business administration and production planning capabilities; and (d) supporting technology upgrading on the farm through information provision and access to finance to upgrade warehouses, cold storage, trellises, and irrigation systems, together with the generation and scaling of useful DT solutions (Sabel and Ghezzi 2020; World Bank 2016). Larger downstream wholesale/retail buyers, including local and international supermarkets and exporters, by establishing quality standards and providing contracts against which financiers can prefinance funding for improved seed varieties and fertilizers, can help pull sustainable quality upgrading through the value chain. By significantly improving the benefits of formality through productivity upgrading support, this type of program provides a better approach to helping informal firms than the traditional approach of lowering the costs of formality.[42] Ideally, both programs should also support the use of DTs not only for value-chain-specific financing and management tools but also for reducing the programs' implementation costs (for example, per-recipient costs of consulting and monitoring could be reduced significantly through appropriate DT solutions).[43]

The SME Business Training and Coaching Loop (in ADEPME), the Programme National de Mise à Niveau des Entreprises (in BMN), and La Fabrique des Champions (in the Delegation for Rapid Entrepreneurship, DER) are examples of instruments that could be used to enhance technology adoption. Senegal already has in place institutions and instruments that could be used to build a national strategy to provide more information and knowledge to businesses. Yet it would be important to clarify what the capacities of these programs are in terms of resources available, including financial and technical expertise, to achieve substantial results and match potential demand. The FAT survey suggests that a relatively small share of SMEs has benefited from public programs aiming to support technology adoption. It would be critical that these or other complementary programs are coordinated, targeted ecosystems according to their potential, and have in place an effective M&E system to inform their actions.

Access to domestic and external markets

Senegal's GVC participation has transitioned over the past three decades from exporting commodities to limited manufacturing and back to commodities. Even within agribusiness GVCs (agriculture and food processing), Senegal is lagging most peers in the extent and quality of GVC participation.[44] Senegal is among five countries in the world with such a GVC boomerang trajectory between commodities and light manufacturing sectors for the period 1990–2015; the other countries are Botswana, Jamaica, the Democratic People's Republic of Korea, and Nicaragua (World Bank 2019). Even within the commodities segment, Senegal was lagging Ghana, Kenya, and Rwanda in the extent of GVC participation in agriculture GVCs in 2015 because of limited progress since the 1990s (Senegal is just above Côte d'Ivoire's level of participation). A similar finding is observed for food processing GVCs. Currently, Senegalese firms participating in GVCs represent 11 percent of traders but 73 percent of the trade value, illustrating that they are the main drivers of trade patterns.[45] Hence, factors that constrain the capacity of exporting firms to import are likely to reduce their export performance, including their capacity to survive in foreign markets and participate in more sophisticated or technology-intensive GVCs.

A priority could be to support modernizing public border entities other than customs, such as sanitary and phytosanitary agencies or the police, to improve agribusiness GVCs. In Senegal, as in the rest of the world, most import and export clearance delays are likely to originate from these agencies instead of customs, as they tend to have lower capabilities and limited use of DTs. Integrating digital tools for risk management in these agencies as well as digital G2B platforms to process import and export licenses could lower supply chain unreliability. Such measures could be particularly effective in fostering GVC participation in agri-food supply chains where Senegal is lagging most peers despite significant comparative advantage.

More generally, improving access to markets can lead to technological upgrading of existing firms and better and more start-ups. A strengthened export promotion program that targets the provision of knowledge and information to potential and existing exporters could be a second stage of intervention for firms that have benefited from the upgrading of managerial and technological capabilities. These "matching" programs could target formal SMEs and large formal firms. These firms would need to go through a managerial and technological diagnostic to identify their main gaps and establish a plan for accessing specific external markets.

International evidence suggests that providing more information and opportunity to access markets can lead to higher productivity of existing firms and more firms. A randomized experiment showed that improving access to foreign markets for rug producers in Egypt led to an increase in profits, quality, and technical efficiency through learning by exporting (Atkin, Khandelwal, and Osman 2017). One of the relevant channels for this "learning-by-exporting" process is that buyers are passing along both information on how to manufacture high-quality rugs (for example, packing that is not too tight) as well as information on what a high-quality product is (for example, the importance of long-term durability). Additional studies on export promotion policies suggest that providing more information and facilitating matching between domestic firms and external buyers can facilitate adoption, both at the extensive (nonexporting firms that start to export) and the intensive (exporting firms exporting more) margins (Cruz, Lederman, and Zoratto 2018). DTs can also directly facilitate access to markets. Evidence exists that digital platforms, such as e-Bay or the equivalent, can facilitate access of small firms to external markets and benefit unskilled workers (Lendle et al. 2016; Cruz, Milet, and Olarreaga 2020).

Regulations to promote entry and innovation

The recent *Senegal Country Private Sector Diagnostic (CPSD)* highlights the importance of improving the investment climate to enable sectors where the country has comparative advantages to grow faster (IFC 2020). The report's findings suggest that firms continue to operate in a difficult business environment. It argues that the most important obstacles to tackle are the lack of a level-playing field in the business environment, access to finance, energy, and limited digital infrastructure connectivity—all aligned with this book's findings.

The government's Programme de Réforme de l'Environnement des Affaires et de la Compétitivité (PREAC3) (now entering its third phase) and the institutional setup under the Secretary-General of the Presidency (SGPR) provide a strong basis upon which to implement ambitious reforms. The PREAC3 constitutes a strong roadmap for investment climate reforms in Senegal. One challenge that the government may face with such an ambitious plan is the breadth of activities encompassed, and the large number of ministries, administrations, and agencies involved to implement changes. Prioritizing will be required first on the most effective measures, particularly during the post-COVID-19 economic recovery, when governments must deal with many emergencies. In addition, strong ownership of reforms at the top of government will be essential to ensure implementation and coordination. As such, the current institutional setup under the SGPR, with APIX in a supporting and coordinating function, organized around thematic topics, may be an effective way to organize reforms.

Some foundational legal and regulatory frameworks are outdated and require being updated to adapt to the evolving business environment. Several critical reforms are highlighted in the PREAC3 that could help dramatically improve the regulatory environment for business. Some of these priorities have also been identified under the G20 Compact with Africa, which is expected to provide strong support to their implementation. Some of the key priorities include labor regulations, access to land, and competition.

Among these proposed reforms, strengthening the competition regulatory framework at the national level is key, given its current de facto lack of effectiveness. A draft competition law is under development. It is critical that it adopt good international practices that are also compliant with the regional West African Economic and Monetary Union (WAEMU) framework. Once approved and supported by new regulations, the national Competition Commission, which currently is not operational, needs to be relaunched. In addition, clarifying the division of labor and respective competencies between Senegal and WAEMU will be essential. This process will involve regional-level interventions, including peer-to-peer exchange with WAEMU counterparts.

Access to finance

Technology upgrading and entrepreneurship need to be enabled with adequate financing. Senegal has a US$1 billion gap in terms of access to finance for micro, small, and medium enterprises (MSMEs) (SME Finance Forum 2021). It is also the country in the WAEMU subregion where firms report being the most financially constrained (World Bank 2014). Financing is mainly provided by the banking sector, with capital markets and venture capital/private equity being nascent while alternative finance does not exist. Most of the banking sector's resources are allocated to the financing of large companies and the state (and state-owned entities).

Adequate financing also means adequate financial services, including more innovation in product design and more proximity to customers in terms of contact points. EcobankPay is a pilot initiated by Ecobank that seeks to equip thousands of merchants in Senegal with virtual electronic payment terminals (linked to merchants' mobile phones as an app).[46] Merchants' virtual e-payment terminals are then paired with e-wallets that allow QR-code-generated payments. This pilot project facilitates Ecobank credit allocation based on payment records. Despite this initiative and a few others, digital financial solutions are not widespread in Senegal. The lack of a supportive environment severely restricts their development and expansion. These restrictions are related to a poor legal framework and an insufficient digital infrastructure (limited access to broadband plus a low penetration of smartphones).

The main barriers to abundant and diverse/innovative financing are lack of competition in the banking sector, lack of credit infrastructure, and insufficient public interventions:

- *There is insufficient competition in the banking sector.* Without stiff competition in the traditional segments (large companies, state, and state-owned companies), credit institutions have no incentive to venture into the MSME segment and close the financing gap. In Senegal, the lack of competition is a three-pronged issue:[47] (a) tight regulatory barriers between categories of financial actors,[48] (b) lack of alternative forms of financing,[49] and (c) insufficient transparency. As an example of lack of transparency, although interest rates and commissions are often posted at customer contact points of financial service providers, there is no requirement to post them online. Also, the Office of Quality of Financial Services, a government agency in charge of consumer protection and information, does not regularly update its price comparator, which also does not cover electronic money services.

- *The credit infrastructure is insufficient.* First, the Regional WAEMU Credit Information Bureau, created recently, is steadily gaining momentum and now covers 8.2 percent of the adult population, according to the latest *Doing Business.* However, the role of this bureau could be considerably expanded if it could collect information from electronic money issuers as well as alternative data from prepaid bills of large utility companies (mobile telephony, electricity). Access to such information would require, however, a reform to a regional law on the Credit Information Bureau. Second, an effective bankruptcy plan would also be useful for two reasons: (a) it would reduce problems of unfair competition induced by "zombie" firms, and (b) it would offer more favorable prospects of recovery and reduction of losses in the event of bankruptcy to creditors (including primarily the banking system). In Senegal, efforts still need to be made to (a) improve the legal and regulatory framework that governs companies in difficulty, (b) train the insolvency proceedings stakeholders (court attorneys, judges, lawyers, and so on), and (c) make operational a framework for out-of-court resolutions. Given the effect of COVID-19 on nonperforming loan ratios in credit institutions, more vigorous solutions may have to be sought through, for example, the establishment of a defeasance structure. Third, reviving the economy and supporting newly created businesses will also require diversifying the guarantees requested by credit institutions. Senegal is a pioneer in this area with the recent law on the Warehouse Receipt System. However, the government must continue its effort and create a single, centralized, and fully digitalized register of movable and immovable collateral. Finally, as part of the credit infrastructure, the legal and regulatory framework in Senegal is either outdated or incomplete in a number of areas, namely, the banking law and the law on microfinance (adopted more than 10 years ago, they do not allow the development of fintech or networks of agents without agency), the law on the credit bureau (see earlier), capital market regulations (many texts are either missing, such as crowdfunding and minibonds, or else obsolete), venture capital regulations, and regulations on USSD codes (deemed too protective of the interests of mobile operators).

- *Better public policies are needed to boost access to finance.* It is the responsibility of the state to design and implement policies addressing market failures and maximizing crowding-in effects (see box 3.5 for examples, including the strengthening of a digital MSME scoring system including informal firms to extend e-credit based on transaction records). Senegal could improve several other public policies. First, the provision of partial guarantees on loans to MSMEs could benefit from a centralized mechanism (there are still many different guarantee mechanisms among different ministries and agencies) in a single structure—ideally, the Guaranty Fund for Priority Investment (FONGIP), to be professionalized, in particular by putting this entity under

Examples of direct public support for the modernization of financial sector market players

The state can play a driving role in the modernization of market players providing financial services. It could set up matching grants (or another form of financial assistance) to do the following:

- Modernize the firms' accounting ecosystems: Technology is changing the way bookkeeping/accounting activities are conducted worldwide. New forms of accounting include (a) the use of mobile to scan/send invoices (clients and providers), (b) the use of remote data entry centers, (c) automated connection to bank accounts (to regularly update firms' cash flows and expended bank account balances), and (d) the use of all these data to enrich firms' credit scoring/rating and further expand credit. Senegal could reform its "Centres de Gestion Agrées" (light public accounting structures) by embracing these new technological trends and providing technical assistance and financial support for firms to adopt and use these DT solutions.
- Help create a regional crowdfunding platform, sponsored by the regional stock exchange

and supported by the existing credit enabling institutions such as ADEPME, DER, and so on. A presentation of the concept was made by the World Bank to Senegal and Côte d'Ivoire.[a]

- Accelerate the adoption of USSD codes by financial service providers (mainly microfinance companies and fintech companies). An analytical work conducted by the World Bank in 2019 shows that, out of 32 requests for USSD codes, only one was implemented in Senegal.[b]
- Stimulate the expansion of networks of financial service agents in low-density areas by setting up a public-private dialogue aiming at defining a viable business model for cash-in/cash-out agents in rural areas.[c]
- Help microfinance companies acquire a Digital Core Banking System and enable them to catch up on their technological gap (e-wallets, remote account opening, e-credit, connectivity to the regional interoperable payment systems, and so on).[d]
- Incentives for the use of payment terminals at merchants (several options could be considered: tax reduction, lottery, and so on).

a. It is included as a background note to this report. See Gonnet 2020, appendix A.
b. *The Use of USSD Codes in the WAEMU*, World Bank, 2019.
c. See also *Support to Digital Connectivity and Transformation in Senegal* (P171740).
d. Digital Core Banking Systems (D-CBS) are a new generation of CBS allowing the full integration of traditional management information systems with modern technologies (smartphones, GPS, USSD, internet, 3G, blockchain), new products (lending based on e-rating) and new channels of distribution (agent banking, subagent banking).

the supervision of the Central Bank of West African States (BCEAO) and by adopting international best practices of corporate governance. Second, the interest rate subsidy policy is wanting. There is a policy of subsidizing interest rates mainly in the rural/agricultural sectors. However, this policy has two major problems: the subsidized rates are poorly targeted (they indiscriminately benefit borrowers, that is, without income level conditions) and are distributed exclusively by a state bank although it would be more efficient to authorize their distribution by other banks and microfinance companies. Third, the government should consider creating an SME financing mechanism to support high-growth SMEs. Finally, there is a need to design and implement a public policy to support the financing of start-ups. Pre-seed and seed types of funds are rarely 100 percent private initiatives worldwide. An intervention of the government in this segment would be welcome. The creation of a public-private start-up fund should be fast-tracked.

Using DT solutions to support technology adoption and entrepreneurship

Digital technologies could enhance high-potential ecosystems through several channels. For example, DTs could be used to promote better and more training, targeting specific products and skills in agriculture. The diffusion of DTs applied to GBFs could also facilitate the expansion of DT supply provision markets, and therefore, the ability of user firms to afford more specific DTs to support production. The demand for these DTs and the support to use them could also be increased by providing better and more information at lower cost, through simpler DTs. Better access to DT enablers, combined with more efficient procurement processes online, could also be important for expanding their markets.

Businesses will need to adapt to COVID-19 and look for DT solutions to reach customers and suppliers. The COVID-19 pandemic is resulting in faster diffusion and adoption of DTs. Some social distancing and mobility restrictions may stay in place for the foreseeable future for fear of contagion and follow-on waves of infection. DT solutions have been one of the main channels for businesses to adapt to the new reality around the world. For example, in China, large farms and agricultural product distributors have been buying high-tech equipment such as drones in a push to reduce human contact, raising demand for DT start-ups in agriculture (Ye 2020). In West Africa, the crisis might be a trigger for mobile payment growth; the incentive to avoid infection could outweigh existing barriers such as a lack of trust in digital payments.

The increase in the demand for DT solutions for business as part of the COVID-19 responses can generate opportunities for DT-providing firms. A recent survey conducted by the World Bank to analyze the effect of COVID-19 on the private sector suggests that about 24 percent of firms increased their use of digital solutions to sell their products and 11 percent invested in digital solutions as a response to the shock, primarily large and formal businesses (Cirera et al. 2020). The survey includes formal and informal firms with 5 or more employees. The adoption of digital solutions was significantly higher among formal and large firms that also have invested more in new digital equipment. The increase in demand for digital solutions is expected to lead to new opportunities for local digital entrepreneurship ecosystems.

Senegal has been stimulating start-ups and innovation, including digital solutions. The government recently adopted new start-up legislation, which aims to create an attractive environment for start-ups and digital innovation. The law was influenced by the new policy-making process called "Dakar Policy Hackathon," which borrowed from hackers' techniques to identify main bottlenecks in the entrepreneurial ecosystem and propose solutions in the shape of a draft law/manifesto.[50] The next priority is to make the new law operational by adopting its implementing regulations, which may cover the following areas:

– Set up a start-up fund or equivalent early-stage financing mechanism (in line with upcoming regulations on PE/VC to be adopted by WAEMU's Regional Council for Public Savings and Financial Markets.
– Define labeling processes and support structures (incubators, accelerators, and so on).
– Explore the provision of grants/credit for research and development.
– Explore the provision of an attractive tax program for venture capital investors and business angels.

– Support the international development of Senegalese start-ups (participation in trade fairs, business development, and so on).
– Invest in the "Senegal Start-Up Nation" brand and promote the use of ".sn."

Important tax facilitation measures have been taken but are not well known. In addition, it will be critical to help disseminate the measures taken in the tax code through the country's LFI 2020, in complement to the start-up law, to ease constraints on start-ups. These include exemption for the first three years of operation of paying the minimum tax and contribution on salaries (CFCE), removal of the threshold of the minimum tax, simplification of the presumptive turnover tax (CGU) applicable to small businesses, and reduction of registration fees.

Accelerating the digitization of government-to-business (G2B) services is also important. This need has been clearly laid out in the PREAC3, which places e-government digitization at the heart of its priorities, because DT solutions allow for simplification of procedures and saving of time and cost for companies and investors, as well as more transparency and quality of service. In addition to improving the country's attractiveness for investment, speeding up digitization helps limit physical contacts and strengthens the continuity of government services. Informed by private sector diagnostics and business surveys, policy makers should consider payment of taxes and credit information for further prioritizing of digitization.[51] However, it also requires ensuring affordable internet availability to ensure that all firms can avail themselves of these opportunities.

NOTES

1. This chapter is based on the results and analysis from Cirera et al. (2021); Cruz, Torres, and Tran (2020); and Atiyas and Dutz (2021) for micro informal enterprises. The policy discussion incorporates inputs provided by Laurent Corthay and Laurent Gonnet.
2. All economic units of formal and informal sectors with built premises were identified, excluding, for instance, itinerant traders and self-employed people working at their customers' homes. Only farms located around residential areas have been included. Also excluded were units carrying out informal activities in agriculture (family farming), livestock, and fishing (carried out by fishermen not registered).
3. According to ANSD, an enterprise is formal when it uses a standardized system of accounting, largely linked to meeting fiscal obligations. Of the 15.2 percent of economic units that keep written accounts, only 19.7 percent use such a standardized accounting system, with most of these using the SYSCOA (West African Accounting System).
4. Using Brazilian data, Ulyssea (2018) shows that the share of such held-back entrepreneurs is small (11.5 percent) relative to low-productivity survivor or subsistence entrepreneurs and will likely stay small until better wage-earning opportunities are available (52.6 percent), and relative to free-riders or parasites that intentionally avoid paying taxes (35.9 percent). However, sufficiently attractive government programs to support the building of capabilities should lead a share of free-rider firms also to seek formalization to take advantage of the benefits from expansion, in addition to held-back entrepreneurs. Sabel and Ghezzi (2020) argue that high-potential informal enterprises are more prevalent than current theories of development lead us to expect. These enterprises face problems overcoming a "quality hurdle"—so defined because the initial decision to overcome related risks and build capabilities to start meeting the demanding technical and organizational requirements of modern supply chains represents a discontinuous jump away from their current, more autonomous operation in the informal sector. This initial quality hurdle stands in contrast to the more continuous learning and monitoring along the ensuing steps of the quality ladder to ensure that progressively more exacting quality standards are met. This

intermediate group of held-back micro-entrepreneurs could benefit from public support for capacity-building extension services, together with complementary support for association to help socialize the costs of learning. A case study is presented on the upgrading by informal farms in Peru in the production for export of high-valued fresh fruits and vegetables like mangoes, avocados, and asparagus—where, in the near-absence of public support, they benefit from assistance from their buyers (large and mid-size exporters) and local cooperatives. They also summarize evidence about the in-between sector in Tanzania (Ellis, McMillan, and Silver 2018), the vegetable export sector in Madagascar, and the fruit and vegetable sectors in Zimbabwe (Henson, Masakurea, and Boselieb 2005). Sabel, O'Donnell, and O'Connell (forthcoming) present a related case study of how environmentally sustainable dairy farming became an engine of growth in Ireland during the past 15–20 years.

5. Estimates based on the RGE database. It refers to labor productivity, measured as value of sales per worker.

6. Detailed results of comparison across countries are provided by Cirera et al. (2020). The state of Ceará in the northeast of Brazil and Vietnam were selected for comparison because they are the locations where these new technology adoption measures are currently available.

7. Hjort and Poulsen (2019) provide evidence on how fast internet affected employment in Africa the importance of the channels discussed in this chapter, including firm entry and productivity gains through technological upgrading.

8. Only 32 percent of Senegalese firms with 5 or more workers and 16 percent of informal micro firms report using a smartphone, versus 75 percent of firms in Ceará, Brazil (Cirera et al. 2021).

9. Bezzina et al. (2019) describe some recent developments on digital infrastructure in Senegal.

10. The Firm-Level Adoption of Technology survey was implemented in Senegal between August 2019 and February 2020. The sample is nationally representative and includes 1,776 establishments with five or more employees, randomly selected from the 2016 Recensement Général des Entreprises (RGE), provided by the Agence Nationale de Statistique et de la Démographie (ANSD). The universe includes establishments with 5 or more employees in agriculture, manufacturing, and services. The sample was stratified by formality status (formal and informal), region (Dakar, Diourbel, Kaolack, Kolda, Saint-Louis, Thiès, and Ziguinchor), size (small: 5–19; medium: 20–99; and large: 100+ employees), and sector (agriculture, food processing, wearing apparel, other manufacturing, retail and wholesale, land transport, finance, health, and other services). The survey was implemented face to face using computer-assisted personal interviews. The findings, based on the 2017–18 Research ICT Africa data on micro enterprises, are reported in the sixth subsection of chapter 2.

11. ERP refers to Enterprise Resource Planning, a category of business management software—typically a suite of integrated applications—that an organization can use to collect, store, manage, and interpret data from many business activities.

12. Cirera et al. (2020) provide a detailed description of all technologies associated with each GBF.

13. For each business function, measures have been developed for the extensive and the intensive margins of technology adoption. For the extensive margin, the firm indicates if each technology is being used to perform a specified business function or not (yes or no question). In case a firm indicates that it uses more than one technology to perform a specific business function, the firm indicates which technology is the most frequently used to perform that task (intensive margin). The variation between the extensive and the intensive margin of adoption at the firm level determines how heterogeneous the adoption of technologies is across business functions within a firm.

14. The technology index, developed by Cirera et al. (2020), summarizes the main indicators of the FAT survey by business function.

15. For the extensive margin, which combines more than one technology, we consider the relationships between the technologies in the functions covered in the FAT survey through four different structures. These structures together describe the relationship between complementarity and substitutability, which are taken into consideration with the index.

16. The fourth industrial revolution, or Industry 4.0, is characterized by the adoption of cyber-physical systems such as robotics and drones, 3D printing, artificial intelligence (AI),

machine learning, and the internet of things across all sectors of the economy. It is reshaping both the way in which and where manufacturing is done.

17. Input testing refers to the process used for selecting and testing the quality of inputs for processed food.

18. The "extensive margin" answers the question of whether an enterprise makes any use at all of a particular technology, while the "intensive margin" answers the question of whether a specific technology is the most frequently used for any particular business function.

19. Industry 4.0 is a term widely used to refer to a group of advanced technologies with a high level of autonomy, such as artificial AI, robots, 3D printers, and cloud computing.

20. Cirera et al. (2020) provide detailed results and estimates of the association between technology adoption and productivity at the subnational level.

21. Results are based on Cirera et al. (2021), controlling for sector and region fixed effects, size, and informality. Technology adoption is strongly and positively associated with size and negatively associated with informality.

22. Results are based on linear regressions to analyze the statistical association between the level of technology adoption and observed obstacles, while controlling for the size of the firms, formality, sector, and region (Cirera et al. 2021).

23. Lack of information refers to issues related to the availability of technologies that could be suitable for the firm, while lack of knowledge refers to issues regarding how to acquire the technology.

24. Atkin et al. (2017) suggest buyers can be an important source of information for technology adoption.

25. This section draws on Atiyas and Dutz (2021).

26. The RIA After Access Business Survey was administered in 2017–18 in Ghana, Nigeria, and Senegal in western SSA; Kenya, Rwanda, Tanzania, and Uganda in eastern SSA; and Mozambique and South Africa in southern SSA. See table C.1 in appendix C for summary statistics comparing FAT and RIA data on enterprises for Senegal.

27. This definition of informality—namely, not complying with relevant laws and regulations—is also the definition used by Ulyssea (2020) in his review of the literature on informality, its causes, and its consequences for development. He further distinguishes between these firms, which he defines as informal on the extensive margin, and formal firms that hire informal workers (without a formal labor contract), which he defines as informal on the intensive margin.

28. Out of 517 firms, 25 use inventory control software and 33 use accounting software, versus 83 businesses that use a smartphone.

29. The annual national poverty line per person according to the 2018–19 household survey is set at CFAF 333,440.5, while the extreme poverty line is set at CFAF 186,869. Converted to monthly levels, the respective moderate and extreme poverty lines are CFAF 27,787 and CFAF 15,572. Because the RIA survey was implemented in fall 2018, no deflation or inflation of these latest poverty lines by the annual CPI is required.

30. Standardized accounting records are kept by 2.9 percent of establishments; 8.8 percent have a NINEA number (tax ID); 2.5 percent are registered at IPRES (Senegal's retirement benefit institution); 2.3 percent are registered at CSS (social security fund); and 12.7 percent have a Registre de Commerce number (trade registry).

31. Based on latest year available for the World Bank Enterprise Survey.

32. For more details, see Cruz, Torres, and Tran (2020).

33. Cruz, Torres, and Tran (2020) use firm-level census data that identify entrepreneurship ecosystems and their potential by combining firm diversity and quality metrics that are geographically concentrated.

34. Following Felkner and Townsend (2011), the measure of statistical significance is Moran's I.A subsector in the algorithm is a 4-digit NAEMA (Nomenclature d'activités des états membres d'AFRISTAT, Observatoire Économique et Statistique d'Afrique Subsaharienne) subsector.

35. For more details about the methodology, which is built on Felkner and Townsend 2011, see Cruz, Torres, and Tran 2020.

36. The tourism value chain includes food preparation services (NAEMA 55), accommodation (NAEMA 56), travel agencies (NAEMA 79), air transportation (NAEMA 51), entertainment activities (NAEMA 90), gambling (NAEMA 92), and recreation activities (NAEMA 93).

37. These procedures align with three key principles for designing effective productivity support policies and institutions emphasized by Dutz (2018): (a) greater transparency in policy

design and priority-setting, identifying the market failures that policies seek to address and minimizing the risks of government failures; (b) greater monitoring and contestability of policies based on rigorous evidence of impact; and (c) effective coordination, both within and between government departments, and between government and business.

38. See Deutschmann, Bernard, and Yameogo (2020) for the effect of a new contracting arrangement in the Senegal groundnut value chain. In partnership with two farming cooperatives in the groundnut basin, they randomly offered micro smallholder farmers across 40 villages a contract providing a bundle of credit to purchase a quality-improving technology (bio-control product Aflasafe SN-01, a new crop treatment to prevent aflatoxin from developing on crops that received regulatory approval and was launched for commercial sale in 2019), training on how to use the technology, and market access (in the form of a guaranteed price premium conditional on quality certification). The average treatment effect was 79 percentage points (uptake was 89 percent in villages where farmers received the contract offer relative to 10 percent in control villages). Farmers in high-risk areas were 49 percent more likely to comply with the strictest international standards. And treatment farmers increased total output sales to the cooperative by about 65 percent. See Deutschmann et al. (2019) for a related study on the impact of bundling of skills and technology (training on improved farming practices), financing (input loans), and crop insurance by the One Acre Fund's program for smallholder farmers in Kenya. Relaxing these multiple productivity constraints simultaneously caused statistically and economically significant increases in maize production by 24 percent and profits by 16 percent.

39. See Ghezzi (2017), who emphasizes three main prerequisites for a successful ME: (a) a private sector capable and interested in problem solving; (b) a public sector willing to participate and able to deliver; and (c) some convener very high up in government capable of inducing cooperation among the different stakeholders, resolving disputes, enacting regulations, and allocating budget.

40. See Sabel and Jordan (2015) for a detailed examination of the Performance Management and Delivery Unit (PEMANDU), an institutional innovation for making, monitoring, and revising ambitious plans for reform involving coordination between public and private actors and among government entities. Goals are translated into key performance indicators (KPIs). Progress is monitored in a regular cycle of meetings across departments, agencies, and (at times) entities from the private sector or civil society. This monitoring reveals coordination problems or flaws in the initial goals, diagnoses their causes, and focuses efforts on solutions. If participants hoard information or reach a deadlock, disputes are "bumped up" to successively higher review bodies. If the deadlock continues, control of the situation passes to superior authorities, with results that may well make all participants worse off. This inflicts what the authors call a "penalty default" and incentivizes the avoidance of deadlocks.

41. For evidence of the effects on improving managerial capabilities and the organization of the firm, see Atkin et al. 2017; Bloom et al. 2013, 2018; Bruhn, Karlan, and Schoar 2018; Cruz, Bussolo, and Iacovone 2018; and Iacovone, Maloney, and McKenzie 2019.

42. According to Ulyssea (2020), the results available in the literature indicate that lowering the costs of formality is not an effective policy to reduce informality. See also Floridi et al. (2020) for a meta-analysis of formalization interventions, concluding that there is some indication that policies increasing benefits are associated with increased formalization rates, but the evidence base is thin, suggesting that further piloting and experimenting are needed to achieve large-scale formalization of the informal economy. They argue that policy makers should focus on strengthening the existing links between the formal and informal economy. ILO (2020) also includes support to informal enterprises to improve their productivity as one of its recommendations. It is an open question whether a more phased approach is preferable, to support informal firms to upgrade their productivity in exchange for formalization over time, or as a quid pro quo for formalization. One possible approach may be to offer significant benefits to informal firms by means of public support and better access to markets in exchange for immediate formalization, with the understanding that they will be shielded from any costs, including taxes and any attendant harassment or other burdens, for a transition period, such as for the first three years; this time could allow any possible regulatory problems associated with excessive licensing requirements as well as with harassment or corruption to be addressed.

43. Further work is needed to understand why there are only a relatively small number of apps specifically designed to help low-skilled, low-income informal farmers and other micro

enterprises. Possible reasons are (a) insufficient demand (due to affordability, illiteracy, local language availability, lack of information and excessive perceived risk of adopting untried products, skill-related problems, or lack of pressure and expected benefits from markets constraining adoption and use of smartphones and apps that ride on these devices for productivity-enhancing purposes); (b) insufficient supply of skilled entrepreneurs able to address this market (or other supply-side barriers linked to digital ID and geo-location mapping to facilitate identification of potential users and their location, and mechanisms to aggregate dispersed individual demands); and (c) a combination of both demand- and supply-side factors. More work is needed to find out what the most appropriate public policy response should be to stimulate the development of this market, if any.

44. Countries' GVC participation is defined as the share of GVC exports in total international exports (World Bank 2019). GVC exports include transactions in which a country's exports embody value added that it previously imported from abroad (backward GVC participation), as well as transactions in which a country's exports are not fully absorbed in the importing country and instead are embodied in the importing country's exports to third countries (forward GVC participation).

45. GVC participation at the firm level is defined as those that import and export (World Bank 2019). Following the fragmentation of production across the world, import and export activities have grown increasingly intertwined. Inputs are typically imported and incorporated in the production of final goods or transformed and exported to other countries, where they may further enter as intermediate inputs in exports to third countries.

46. See Ndiaye (2020) for a case study on EcobankPay.

47. Unlike other countries, Senegal does not suffer from an excessive concentration of credit supply, with the three largest banks holding only 35 percent of the market share.

48. An important example of barriers is that because of inadequate regulation, banks and microfinance companies cannot participate in the e-wallets market. This, de facto, confers a monopoly to mobile operators.

49. Capital markets play an almost insignificant role in the financing of large companies and the financing of projects. Similarly, unlike other countries in SSA, there are no crowdfunding platforms in Senegal. Factoring and leasing are also underdeveloped forms of financing.

50. The policy hackathon involved more than 50 actors of the ecosystem in situ, and more than 200 contributors via chatbot. The law touches upon critical dimensions of start-ups' life cycles such as entry, access to finance, access to public procurement, intellectual property rights protection, role of support structures, and so on.

51. On digital payment of taxes, the government has taken several measures aimed at reducing the tax burden on companies' cashflow in the context of the COVID crisis, including partial relief of tax debts, reimbursement of VAT credits, postponement of payment deadlines, and so on. In addition, as part of the Yaatal program, the DGID is accelerating digitization of services, in particular through a new financial statement filing application scheduled to be launched in July, as well as the expansion of the eTax online filing and payment platform, the Mon Espace Perso app for individual and small taxpayers. New mobile filing and payment platforms (mTax) are also in the pipeline. On credit information and contract enforcement, the PREAC3 identifies actions that are aimed at strengthening credit information and securing its regulatory ecosystem. These include (a) the implementation of the electronic collateral registry by operationalizing the RCCM's electronic directory of movable assets, (b) the operationalization of warehouse receipts, (c) the simplification and dematerialization of contract registration rules, and (d) the implementation of the electronic land register and online procedures for real estate transactions.

REFERENCES

Abate, Gashaw Tadesse, Shahidur Rashid, Carlo Borzaga, and Kindie Getnet. 2016. "Rural Finance and Agricultural Technology Adoption in Ethiopia: Does the Institutional Design of Lending Organizations Matter?" World Development 84: 235–53.

Atiyas, Izak, and Mark Dutz. 2021. "Digital Technology Uses Among Informal Micro-Sized Firms: Productivity and Jobs Outcomes in Senegal." Policy Research Working Paper 9573,

World Bank, Washington, DC. https://openknowledge.worldbank.org /handle/10986/35251.

Atkin, David, Azam Chaudhry, Shamyla Chaudry, Amit K. Khandelwal, and Eric Verhoogen. 2017. "Organizational Barriers to Technology Adoption: Evidence from Soccer-Ball Producers in Pakistan." Quarterly Journal of Economics 132 (3): 1101–64.

Audretsch, David, Marcio Cruz, and Jesica Torres. 2020. "Entrepreneurship Ecosystems in Developing Countries." World Bank, Washington, DC.

Bezzina, Jerome, Aneliya Muller, Zaki Badie Khoury, and Mouhamed Tidiane Seck. 2019. Country Diagnostic of Senegal (English). Washington, DC: World Bank.

Bircan, Çağatay, and Ralph De Haas. 2020. "The Limits of Lending? Banks and Technology Adoption Across Russia." Review of Financial Studies 33 (2): 536–609.

Bloom, Nicholas, Benn Eifert, Aprajit Mahajan, David McKenzie, and John Roberts. 2013. "Does Management Matter? Evidence from India." Quarterly Journal of Economics 128 (1): 1–51.

Bloom, Nicholas, Aprajit Mahajan, David McKenzie, and John Roberts. 2018. "Do Management Interventions Last?: Evidence from an Experiment in India." Policy Research Working Paper 8339, World Bank, Washington, DC. http://documents.worldbank.org/curated /en/346001518549616744/Do-management-interventions-last-evidence-from-India.

Bruhn, Miriam, Dean Karlan, and Antoinette Schoar. 2018. "The Impact of Consulting Services on Small and Medium Enterprises: Evidence from a Randomized Trial in Mexico." Journal of Political Economy 126 (2): 635–87.

Caselli, Francesco, and Wilbur John Coleman. 2001. "Cross-Country Technology Diffusion: The Case of Computers." American Economic Review 91 (2): 328–35.

Cirera, Xavier, Diego Comin, and Marcio Cruz. 2020. "A New Approach to Measure Technology Adoption: The Firm-Level Adoption of Technology (FAT) Survey." Internal report, World Bank, Washington, DC.

Cirera, Xavier, Marcio Cruz, Diego Comin, and Kyung Min Lee. 2020. "Technology Within and Across Firms," NBER Working Paper 28080, National Bureau of Economic Research, Cambridge, MA.

Cirera, Xavier, Marcio Cruz, Diego Comin, and Kyung Min Lee. 2021. "Firm-Level Adoption of Technologies in Senegal." Policy Research Working Paper 9657, World Bank, Washington, DC.

Cirera, Xavier, Marcio Cruz, Leonardo Iacovone, and Jesica Torres. 2020. "Quantifying the Impact of COVID-19 on the Private Sector in Senegal." Internal report, World Bank, Washington, DC.

Cole, Harold L., Jeremy Greenwood, and Juan M. Sanchez. 2016. "Why Doesn't Technology Flow from Rich to Poor Countries?" Econometrica 84 (4): 1477–521.

Comin, Diego, and Bart Hobijn. 2010. "An Exploration of Technology Diffusion." American Economic Review 100 (5): 2031–59.

Comin, Diego, and Martí Mestieri. 2018. "If Technology Has Arrived Everywhere, Why Has Income Diverged?" American Economic Journal: Macroeconomics 10 (3): 137–78.

Cruz, Marcio, Maurizio Bussolo, and Leonardo Iacovone. 2018. "Organizing Knowledge to Compete." Journal of International Economics 111 (C): 1–20.

Cruz, Marcio, Daniel Lederman, and Laura De Castro Zoratto. 2018. "Anatomy and Impact of Export Promotion Agencies." In Research Handbook on Economic Diplomacy, edited by Peter A. G. van Bergeijk and Selwyn J. V. Moons. London: Edward Elgar.

Cruz, Marcio, Emmanuel Milet, Marcelo Olarreaga. 2020. "Online Exports and the Skilled-Unskilled Wage Gap." PLoS ONE 15 (5): e0232396. https://doi.org/10.1371/journal .pone.0232396.

Cruz, Marcio, Jesica Torres, and Trang Tran. 2020. "Entrepreneurship Ecosystems in Senegal: Challenges and Opportunities of Digital Technologies." World Bank, Washington, DC.

Dalberg Group. 2020. "Case Study on the CommAgri and Commango Apps in Senegal." Internal report, World Bank, Washington, DC.

Deutschmann, Joshua W., Maya Duru, Kim Siegal, and Emilia Tjernström. 2019. "Can Smallholder Extension Transform African Agriculture?" National Bureau of Economic Research Working Paper 26054, Cambridge, MA.

Deutschmann, Joshua W., Tanguy Bernard, and Ouambi Yameogo. 2020. "Contracting and Quality Upgrading: Evidence from an Experiment in Senegal." University of Wisconsin–Madison, November 15. https://jwdeutschmann.com/research/jmp/.

Dutz, Mark A. 2018. Jobs and Growth: Brazil's Productivity Agenda. International Development in Focus. Washington, DC: World Bank. https://openknowledge.worldbank.org/handle/10986/298008.

Dutz, Mark A., Rita Almeida, and Truman Packard. 2018. The Jobs of Tomorrow: Technology, Productivity, and Prosperity in Latin America and the Caribbean. Washington, DC: World Bank. https://openknowledge.worldbank.org/handle/10986/29617.

Ellis, Mia, Margaret McMillan, and Jed Silver. 2018. "Employment and Productivity Growth in Tanzania's Service Sector." In Industries without Smokestacks: Industrialization in Africa Reconsidered, edited by Richard S. Newfarmer, John Page, and Finn Tarp. Oxford: Oxford University Press.

Felkner, John S., and Robert M. Townsend. 2011. "The Geographic Concentration of Enterprise in Developing Countries." Quarterly Journal of Economics 126 (4): 2005–61.

Floridi, Andrea, Binyam Afewerk Demena, and Natascha Wagner. 2020. "Shedding Light on the Shadows of Informality: A Meta-analysis of Formalization Interventions Targeted at Informal Firms." Labour Economics 67. https://www.sciencedirect.com/science/article/pii/S0927537120301299?via%3Dihub.

Ghezzi, Piero. 2017. "Mesas Ejecutivas in Peru: Lessons for Productive Development Policies." Global Policy 8 (3): 369–80. https://doi.org/10.1111/1758-5899.12457.

Henson, Spencer, Oliver Masakurea, and David Boselieb. 2005. "Private Food Safety and Quality Standards for Fresh Produce Exporters: The Case of Hortico Agrisystems, Zimbabwe." Food Policy 30 (4): 371–84.

Hjort, Jonas, and Jonas Poulsen. 2019. "The Arrival of Fast Internet and Employment in Africa." American Economic Review 109 (3): 1032–79.

Iacovone, Leonardo, William Maloney, and David McKenzie. 2019. "Improving Management with Individual and Group-Based Consulting: Results from a Randomized Experiment in Colombia." Policy Research Working Paper 8854, World Bank, Washington, DC.

IFC (International Finance Corporation). 2020. Creating Markets in Senegal: Country Private Sector Diagnostic. Washington, DC: IFC.

ILO (International Labour Organization). 2020. Diagnostic de l'économie informelle au Sénégal. Geneva: ILO.

Lendle, Andreas, Marcelo Olarreaga, Simon Schropp, and Pierre-Louis Vézina. 2016. 'There Goes Gravity: eBay and the Death of Distance." Economic Journal 126 (591): 406–41.

Midrigan, Virgiliu, and Daniel Yi Xu. 2014. "Finance and Misallocation: Evidence from Plant-Level Data." American Economic Review 104 (2): 422–58.

Ndiaye, Omar. 2020. "Analyse de EcobankPay au Sénégal." Case study, World Bank, Washington, DC.

Sabel, Charles F., and Piero Ghezzi. 2020. "The Quality Hurdle: Toward a Development Model That Is No Longer Industry-centric." Informal publication, November 5. http://www2.law.columbia.edu/sabel/papers/QualityHurdle_Nov-6-2020.pdf.

Sabel, Charles F., and Luke Jordan. 2015. "Doing, Learning, Being: Some Lessons Learned from Malaysia's National Transformation Program." Competitive Industries and Innovation Program, World Bank, Washington, DC. https://www.theciip.org/sites/ciip/files/documents/PEMANDU%20Study%20--Final.pdf.

Sabel, Charles F., Rory O'Donnell, and Larry O'Connell. Forthcoming. "Self-Organization under Deliberate Direction: Irish Dairy and the Possibilities of a New Climate Change Regime." Under review by Regulation and Governance. http://www2.law.columbia.edu/sabel/papers/Self%20Organisation%20under%20Deliberate%20Direction_Sabel-O'Donnell-O'Connell_15%20Feb%202019.pdf.

SME Finance Forum. 2021. "MSME Finance Gap." Managed by IFC, World Bank Group. https://www.smefinanceforum.org/data-sites/msme-finance-gap.

Ulyssea, Gabriel. 2018. "Firms, Informality, and Development: Theory and Evidence from Brazil." American Economic Review 108 (8): 2015–47.

Ulyssea, Gabriel. 2020. "Informality: Causes and Consequences for Development." Annual Review of Economics 12: 525–46.

UNECE (UN Economic and Social Council). 2016. "A Road Toward Paperless Trade: Senegal's Experience." Case Stories, Trade Facilitation Implementation Guide, United Nations, New York.

Vijil, Mariana. 2020. "Supply Chain Reliability Matters: The Role of Import Uncertainty on Export Performance in Senegal." Internal report, World Bank, Washington, DC.

Vijil, Mariana, Laurent Wagner, and Martha Tesfaye Woldemichael. 2019. "Import Uncertainty and Export Dynamics." Policy Research Working Paper 8793, World Bank, Washington, DC.

World Bank. 2014. Senegal Enterprise Survey. Washington, DC: World Bank. https://microdata .worldbank.org/index.php/catalog/2262.

World Bank. 2016. World Development Report 2016: Digital Dividends. Washington, DC: World Bank.

World Bank. 2018. Senegal: Systematic Country Diagnostic. Washington, DC: World Bank Group.

World Bank. 2019. World Development Report 2020: Trading for Development in the Age of Global Value Chains. Washington, DC: World Bank.

World Bank. 2020. "Trade and COVID-19 Guidance Note: Managing Risk and Facilitating Trade in the COVID-19 Pandemic." World Bank, Washington, DC.

Ye, Wendy. 2020. "Chinese Agriculture Drone Makers See Demand Rise Amid Coronavirus Outbreak." CNBC, March 9, 2020. https://www.cnbc.com/2020/03/10/chinese-agriculture -drone-makers-see-demand-rise-amid-coronavirus-outbreak.html.

APPENDIX A

Background Studies

CHAPTER 1

Calderón, César, and Catalina Cantú. 2020. "Impact of Digital Economy on Growth and Poverty Reduction." Internal report, World Bank, Washington, DC.

Cirera, Xavier, Marcio Cruz, Leonardo Iacovone, and Jesica Torres. 2020. "Quantifying the Impact of COVID-19 on the Private Sector in Senegal." Internal report, World Bank, Washington, DC.

Porto, Guido. 2020. "Digital Technologies and Poorer Households' Income Earning Choices in Sub-Saharan Africa: Analytical Framework and a Case Study for Senegal." Internal report, May 20, World Bank, Washington, DC.

République du Sénégal, Ministère de l'Economie, des Finances et du Plan. 2019. "Plan Sénégal Émergent: Plan d'Actions Prioritaires 2019–2023." Dakar, Sénégal.

République du Sénégal, Ministère des Postes et des Télécommunications. 2016. "Stratégie Sénégal Numérique 2016–2025." Dakar.

Rodríguez-Castelán, Carlos, Samantha Lach, Takaaki Masaki, and Rogelio Granguillhome Ochoa. 2021. "How Do Digital Technologies Affect Household Welfare in Developing Countries? Evidence from Senegal." Policy Research Working Paper 9576, World Bank, Washington, DC.

CHAPTER 2

Atiyas, Izak, and Toker Doğanoğlu. 2020. "Using the RIA Data Set to Explore Correlates of Mobile Internet Use in Senegal." Internal report, World Bank, Washington, DC.

Enamorado, Ted, Takaaki Masaki, Carlos Rodríguez-Castelán, and Hernan Winkler. 2020. "Local Welfare Effects of Digital Technologies in Senegal." Internal report, World Bank, Washington, DC.

Masaki, Takaaki, Rogelio Granguillhome Ochoa, and Carlos Rodríguez-Castelán. 2020. "Broadband Internet and Household Welfare in Senegal." Policy Research Working Paper 9386, World Bank, Washington, DC.

Oughton, Edward. 2020. "Policy Options for Affordable Digital Infrastructure Expansion: A Simulation Model for National and Regional Markets in Africa." Internal report, World Bank, Washington, DC.

Rodríguez-Castelán, Carlos, Rogelio Granguillhome Ochoa, Samantha Lach, and Takaaki Masaki. 2021. "Mobile Internet Adoption in West Africa." Policy Research Working Paper 9560, World Bank, Washington, DC.

CHAPTER 3

Atiyas, Izak, and Mark Dutz. 2021. "Digital Technology Uses Among Informal Micro-Sized Firms: Productivity and Jobs Outcomes in Senegal." Policy Research Working Paper 9573, World Bank, Washington, DC.

Cirera, Xavier, Marcio Cruz, Diego Comin, and Kyung Min Lee. 2021. "Firm-Level Adoption of Technologies in Senegal." Policy Research Working Paper 9657, World Bank, Washington, DC.

Cruz, Marcio, Jesica Torres, and Trang Tran. 2020. "Entrepreneurship Ecosystems in Senegal: Challenges and Opportunities of Digital Technologies." Internal report, World Bank, Washington, DC.

Dalberg Group. 2020. "Case Study on the CommAgri and Commango Apps in Senegal." Internal report, World Bank, Washington, DC.

Gonnet, Laurent. 2020. "Digital Marketplace Platform for MSME Loans." Slides, World Bank, Washington, DC.

Ndiaye, Omar. 2020. "Analyse de EcobankPay au Sénégal." Case study, World Bank, Washington, DC.

Vijil, Mariana. 2020. "Supply Chain Reliability Matters: The Role of Import Uncertainty on Export Performance in Senegal." Internal report, World Bank, Washington, DC.

Firm-Level Adoption of Technology Survey

The FAT is organized in five modules:

- Module A—Collects general information about the characteristics of the establishment.
- Module B—Covers the technologies used in eight generic business functions.
- Module C—Covers the use of technologies for functions that are specific to each of 10 industry and services sectors.
- Module D—Includes questions about the drivers and barriers for technology adoption.
- Module E—Collects information on employment, balance sheet, and performance, which allows computation of labor productivity and other measures at the company level.

The survey differentiates between general business functions that all firms conduct regardless of the sector where they operate (for example, businesses' administration-related human resources and finance, production planning, sourcing and procurement, sales, method of payment) and sector-specific functions/production processes that are relevant only for companies in a given sector (for example, food refrigeration in food processing, or sewing in apparel). Information about technologies used in the former is collected in module B, while information on sector-specific technologies is collected in module C. To design modules B and C, the survey draws upon the knowledge of experts in production and technology in various fields and sectors.

A detailed description of the FAT survey and the technology adoption index are described by Cirera et al. (2021). Figure B.1 provides an example of the index in the extensive and intensive margins for one general business function (left) and one sector-specific function (right), following a vertical quality ladder. The example suggests that this firm performs administrative processes associated with human resources, financing, and accounting through handwritten processes and computers with standard software, but the most frequently used method is handwritten processes. In this case, the extensive margin index is 2, while the intensive margin is 1. For storage, the firm uses the most basic technology, and the index is 1 for both extensive and intensive margins.

FIGURE B.1

Example of the technology index

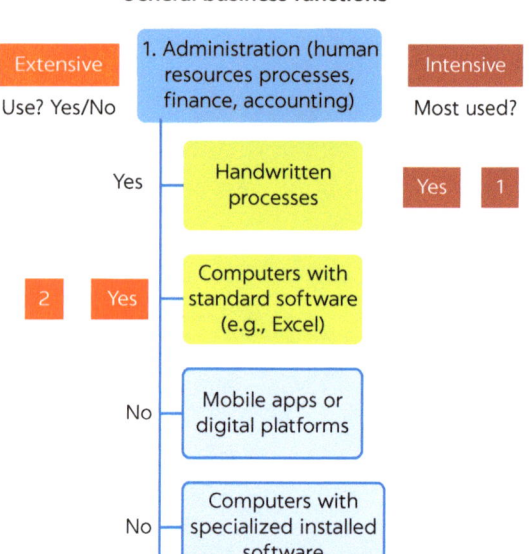

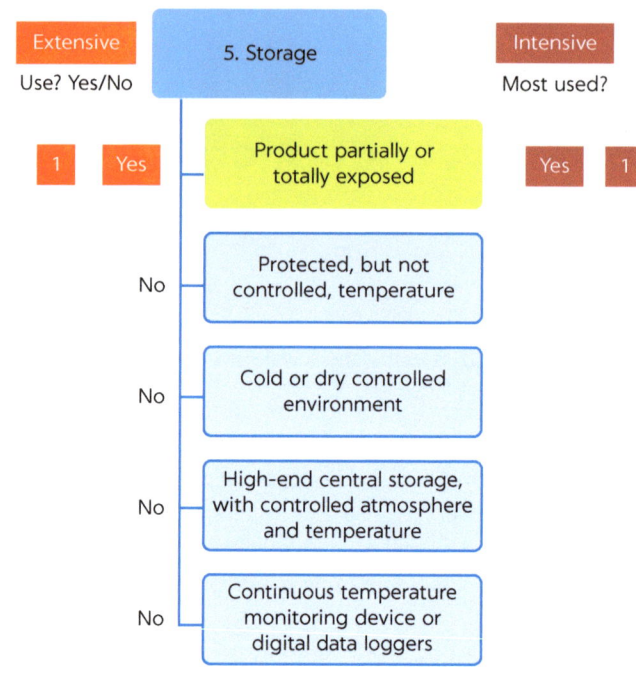

Source: Cirera et al. 2021.

REFERENCE

Cirera, Xavier, Marcio Cruz, Diego Comin, and Kyung Min Lee. 2021. "Firm-Level Adoption of Technologies in Senegal." Policy Research Working Paper 9657, World Bank, Washington, DC.

Summary Statistics of RIA and FAT Data on Enterprises

TABLE C.1 **Firm characteristics—means (medians in parentheses)**

	SAMPLE SECTOR SHARES (%)			FIRM SIZE (FULL-TIME WORKERS)			FIRM AGE (YEARS)		
	RIA-SEN	RIA-SSA	FAT	RIA-SEN	RIA-SSA	FAT	RIA-SEN	RIA-SSA	FAT
ALL FIRMS [total no. of firms]	[517]	[3,889]	[1,786]	1.3 (0)	0.7 (0)	29.0 (6)	8.0 (6)	10.1 (7)	16.9 (14)
Agriculture	13	10	11	0.7 (0)	0.8 (0)	39.2 (7)	8.4 (6)	12.9 (8)	20.5 (19)
Manufacturing	3	5	38	2.0 (1)	1.3 (0)	35.1 (7)	7.1 (5)	12.8 (9)	15.9 (14)
Trade (wholesale/retail)	57	63	21	0.8 (0)	0.5 (0)	19.1 (6)	7.9 (6)	9.7 (6)	16.8 (14)
Other services	27	22	30	2.6 (1)	1.1 (0)	24.3 (10)	8.5 (6)	10.5 (7)	17.1 (14)

	MANAGER'S EDUCATION (YEARS)			FIRMS WITH FEMALE OWNER (%)			FIRMS WITH ELECTRICITY (%)		
	RIA-SEN	RIA-SSA	FAT	RIA-SEN	RIA-SSA	FAT	RIA-SEN	RIA-SSA	FAT
ALL FIRMS	4.8 (0)	7.4 (6)	10.7 (6)	35	51	28	64	54	82
Agriculture	5.2 (0)	7.1 (6)	8.1 (6)	56	50	43	34	28	35
Manufacturing	4.7 (6)	7.5 (6)	9.0 (6)	29	39	20	76	47	92
Trade	5.0 (0)	7.0 (6)	11.6 (9)	34	54	33	64	53	86
Other services	4.4 (0)	8.5 (6)	13.1 (16)	28	43	31	80	71	85

Source: Atiyas and Dutz 2021.
Note: The tables report totals, shares, and means based on unweighted data; medians are reported in parentheses. RIA covers the data from the RIA After Access Business Survey compiled by Research ICT Africa (RIA) in 2017–18. RIA-SSA covers the other eight SSA countries. FAT covers the 2019–20 Senegal Firm-Level Adoption of Technology survey. Brackets contain the total number of firms in each of the samples. Sample sector shares for RIA firms are based on a larger total number of responses, as firms are not asked their primary activity but rather separately whether they produce agricultural products or manufacturing products, whether they trade or sell goods, and whether they are providing another service (with a total of 603 responses for Senegal, 4,321 for non-Senegal SSA). Manager's education is measured in years based on: "What is the highest level of education of the business manager?" The coding is as follows: "None" = 0, Primary = 6, "Secondary" = 13, "Tertiary: Diploma /Certificate" = 15, "Tertiary: Bachelor's" = 16, "Tertiary: Master's" = 19. Firms with electricity are based on a yes/no answer to "Do the business premises have electricity?" FAT = Firm-Level Adoption of Technology survey; RIA = Research ICT Africa; SEN = Senegal; SSA = Sub-Saharan Africa.

Use of DTs by Micro Informal Enterprises

TABLE D.1 **Use of DTs by micro informal enterprises, by age and gender**

| | ACCESS TECHNOLOGIES | | | | EXTERNAL-TO-FIRM TRANSACTION | | | | | | | | | INTERNAL-TO-FIRM | | |
| | | | | | UPSTREAM AND DOWNSTREAM PRODUCTS | | | | | FINANCE | LABOR | GOVERNMENT | | MANAGEMENT | | WORKERS |
	USE ANY MO-BILE	USE SMART-PHONE	USE COM-PUTER	HAVE WEB-SITE	LOOK FOR SUP-PLIERS	PAY SUP-PLIERS	UNDER-STAND CUS-TOMERS	USE E-COMMERCE	RECEIVE PAY-MENTS	USE ONLINE BANK-ING	RECRUIT WORKERS	INTERACT W/ GOVT.	PAY TAX-ES	ACCOUNTING SOFTWARE	INVENTORY/ POINT-OF-SALE SOFTWARE	PAY WORKERS
All firms	89.4	18.3	9.0	4.6	5.8	25.0	12.8	8.1	25.9	4.1	2.2	1.7	4.2	6.8	5.3	6.1
Youth	92.1	27.1	12.0	7.9	10.4	27.7	22.2	11.6	30.6	9.0	3.1	2.2	5.4	10.9	7.5	8.8
Older	88.6	14.1	7.9	3.2	3.9	23.8	8.2	6.7	22.9	1.9	1.3	1.5	3.8	5.0	4.4	5.1
Women	83.5	15.5	8.3	3.6	5.6	19.9	11.9	6.6	20.3	3.8	2.2	2.0	3.8	7.1	6.4	3.6
Men	92.6	19.8	9.4	5.1	5.9	28.7	13.3	8.9	29.0	4.2	2.2	1.6	4.5	6.6	4.7	7.5
Younger women	88	27.4	13.1	7.0	9.6	25.7	23.7	9.1	25.0	6.7	3.9	3.0	3.7	13.1	12.4	4.2
Older women	82.1	10.1	6.2	2.0	3.8	14.7	6.4	5.6	16.9	2.4	1.5	1.5	4.0	4.4	3.7	3.4
Younger men	94.4	26.9	11.4	8.3	10.9	28.8	21.4	13.0	33.8	10.3	2.6	1.7	6.4	9.6	4.6	11.4
Older men	91.9	16.2	8.7	3.8	3.9	28.5	9.1	7.3	26.0	1.6	1.2	1.5	3.7	5.4	4.8	5.9

Source: Atiyas and Dutz 2021.

Note: All responses are shares (%) of firms based on weighted data. *Use any mobile* is in response to "Does the business manager have a mobile?" irrespective of whether it is for private use, business use, or both. *Smartphone users, Use s-phone,* answered "yes" to "How does the business access the internet: Mobile broadband (3G/4G, wireless)?" *Use computer* is a nonzero response to "How many computers does your business have?" *Have website* is in response to "Does your business have a website?" Reported answers to "What do you use the internet for?" include *Look for suppliers* (online), *Use e-commerce* (selling products and services online), *Use online banking, Recruit workers,* and *Interact with government.* Reported answers to "Does the business use mobile money for . . ." include *Pay Suppliers, Receive payments, Pay taxes,* and *Pay workers. Understand customers* is an "agree" (as opposed to "not sure" or "disagree") response to the question "Regarding the internet/social media use, it helps to understand our customers better." The management-related questions are "Does your company use accounting software?" and "Does your company make use of inventory control/point of sale (POS) software?" (both asked in the computer section of the questionnaire).

TABLE D.2 **Productivity, profits, and exporting outcomes**

	LABOR PRODUCTIVITY		TOTAL PROFITS		SHARE OF EXPORTERS	
	DT USERS	NONUSERS	DT USERS	NONUSERS	DT USERS	NON-USERS
Use smartphone	635,841 (178,750)	294,474 (90,000)	3,022,483 (421,000)	687,005 (135,000)	37.8	4.5
Look for suppliers	637,036 (195,000)	313,055 (85,000)	5,430,860 (550,000)	784,328 (130,000)	58	7.3
Pay suppliers	459,244 (135,000)	287,984 (73,500)	944,730 (240,000)	1,093,020 (100,000)	16	8.3
Understand customers	637,723 (195,000)	286,974 (82,500)	3,898,527 (475,000)	637,767 (120,000)	45.4	5.1
Use e-commerce	682,669 (295,000)	301,016 (82,500)	3,203,469 (550,000)	865,891 (125,000)	62.3	5.7
Receive payments	513,978 (140,000)	267,985 (74,000)	1,164,008 (320,000)	1,016,851 (100,000)	23.6	5.6
Use online banking	706,544 (446,875)	316,063 (87,500)	9,094,222 (1,270,000)	714,349 (130,000)	77.7	7.4
Recruit workers	596,660 (107,500)	325,914 (87,500)	13,402,555 (510,000)	772,722 (130,000)	73.7	8.8
Interact w government	762,341 (446,875)	324,496 (87,500)	17,467,115 (13,300,000)	770,351 (130,000)	93	8.9
Pay taxes	767,795 (290,000)	312,658 (85,000)	2,049,276 (565,000)	1,011,071 (130,000)	29.5	9.4
Accounting s/w	1,146,152 (896,500)	272,729 (85,000)	8,660,300 (5,855,000)	503,186 (126,000)	53	7.2
Inventory/POS	1,400,435 (1,407,407)	272,315 (85,000)	10,989,778 (6,900,000)	501,856 (126,000)	56.1	7.7
Pay workers	796,354 (193,000)	301,658 (85,000)	2,445,788 (380,000)	964, 478 (127,000)	40.7	8.3

Source: Atiyas and Dutz 2021.
Note: Labor productivity is as value added (total sales minus raw materials & intermediate inputs plus water & electricity used in production) divided by total number of full-time working people including owners. Profits are measured as value added minus salary and wages and water and electricity costs. Exporting reflects shares of firms that report having international customers. Labor productivity and profits are means of monthly values, in local currency (CFAF); medians are reported in parentheses. No use of smartphone represents use of 2G phones rather than no use of any mobile phone.

TABLE D.3 **More and better job outcomes**

	MORE JOBS: FIRM SIZE		BETTER JOBS: AVERAGE WAGES		BETTER JOBS: ENTREPRENEUR PROFITS	
	DT USERS	NONUSERS	DT USERS	NONUSERS	DT USERS	NONUSERS
Use smartphone	4.7 (3)	2.4 (2)	107,545 (30,000)	46,322 (15,000)	2,343,481 (361,000)	657,806 (130,000)
Look for suppliers	6 (3)	2.5 (2)	192,883 (70,000)	47,239 (15,000)	4,089,560 (475,000)	709,335 (125,000)
Pay suppliers	3.4 (2)	2.5 (1)	84,635 (30,000)	50,643 (11,667)	879,218 (219,000)	915,578 (98,500)
Understand customers	5.6 (3)	2.3 (1)	131,351 (33,333)	43,865 (30,000)	2,935,845 (400,000)	608,437 (115,000)
Use e-commerce	4.4 (3)	2.6 (2)	165,870 (38,750)	47,182 (15,000)	2,042,990 (475,000)	806,191 (119,000)
Receive payments	2.9 (3)	2.7 (1)	77,612 (25,000)	53,140 (12,500)	1,071,457 (290,000)	848,392 (90,000)
Use online banking	9.5 (3)	2.5 (2)	58,408 (33,333)	61,815 (15,000)	6,752,918 (1,270,000)	658,496 (126,000)
Recruit workers	14.6 (6)	2.5 (2)	27,693 (33,333)	63,134 (15,000)	9,420,440 (510,000)	711,599 (126,000)
Interact w government	17.5 (21)	2.5 (2)	20,160 (4,500)	62,911 (16,667)	12,249,042 (6,650,000)	709,515 (126,000)
Pay taxes	2.5 (3)	2.8 (2)	32,753 (20,000)	63,346 (15,000)	1,695,832 (510,000)	871,354 (125,000)
Accounting s/w	8.1 (4)	2.7 (2)	130,028 (16,667)	52,498 (16,667)	6,840,849 (1,783,000)	475,646 (120,000)
Inventory/POS	8.7 (4)	2.4 (1)	158,572 (20,000)	51,370 (15,000)	8,694,908 (6,650,000)	472,586 (120,000)
Pay workers	7.1 (3)	2.5 (1)	60,213 (50,000)	61,767 (15,000)	2,283,909 (275,000)	816,566 (126,000)

Source: Atiyas and Dutz 2021.
Note: Firm size is the number of full-time employees plus owners. *Average wages* reflect salary & wages divided by full-time employees. *Entrepreneur profits* are measured as value added minus salary & wages divided by number of owners. Average wages and profits are means of monthly values, in local currency (CFAF); medians are reported in parentheses. No use of smartphone represents use of 2G phones rather than no use of any mobile phone.